The Art of The Builder

Elevating Construction Superintendents

A Principle Based Leadership Guide for Assistant Supers and Superintendents in Construction

Elevating Construction Superintendents

Written by Jason Schroeder

Edited by Joan Willden

Dedication

I want to personally thank all the builders who have been patient with me and this work. The intent of this first edition is to roll out the first lean version of a book that will help our field builders to be more effective, lead better, and lead remarkable lives. If this work helps to do that, it has done its job but will not be complete without continual feedback, added content, and additional stories we hope will be provided by our builders in the field. Without that input, this book will not reach its ultimate goal. This book is dedicated to the leaders in construction who bear the brunt of all the stress, variation, and impossible circumstances they deal with every day. I have always said, "No one will ever truly know stress unless you have been a superintendent." My hope is that this work will make your role a little more manageable, a little more fun, and a little less impactful to your family. On we go!

Table of Contents

Preface by Jason Schroeder

Some of the greatest leaders in history were shaped by the books they read. George Washington was heavily influenced by a book on manners and etiquette that helped him develop a physical presence that few dared to question. Abraham Lincoln transformed himself from a poor, uneducated boy on the frontier to the President of the United States with the help of the texts he had access to in his early years. General George S. Patton, arguably one of the most successful combat generals in American history, developed his skills by reading books about war, history, and military combat strategy.

When I became a superintendent, I wanted to follow the lead of these great minds and sharpen my leadership skills using books related to my field. I could find guides and manuals about project management, field engineering, lean, BIM, and more, but they were all primarily dedicated to function and form. Where were the books about the art of being a builder? Where was the construction equivalent of Sun Tzu's *Art of War*? Because I couldn't find a book that wholly engaged me, I decided to write one myself and aptly name it--*The Art of the Builder*.

The idea for this book emerged from observations spanning seven years during my early career when I worked for a large general contractor as a field trainer. I would travel from project to project across the United States observing, training, and learning lessons through each interaction. I boarded a plane over five hundred times to perform my responsibilities. The principles gleaned from this book will help leaders command with purpose to better control the outcomes of their projects.

This book is styled after the best seller, *How to Win Friends and Influence People*, by Dale Carnegie. The role of a superintendent is too complex and detailed to summarize with just one story so my intent is to communicate principles generally organized into many stories. Additionally, much like Carnegie's classic book, I intend to collect and insert feedback as future revisions are published. The inclusion of stories, principles, and feedback will be critical to representing the role of a superintendent in various circumstances. I invite you to share your insights, added principles, and stories that will enhance this book for future editions.

Additionally, this book is specifically written for assistant superintendents and superintendents in our industry. These positions are typically the Super 1 and 2 positions. The next work in this series will be *Elevating Construction Senior Superintendents*, which will apply to the Super 3, 4, and 5 positions of project superintendent, senior superintendent, and general superintendent. There is much more that could be written in this work but it felt appropriate to write about the concepts that would take this level of superintendent to the next level. If you observe theory or steps missing in this work, it will either be covered in the next book or in-person training and boot camps. It may be information that should be learned as company-specific-training or will be added to a second revision. Most superintendents become naturally skilled with things like cranes, industry norms, and means and methods. That is not something that is typically missing from their toolbox. Our intent is to provide supers with tools that are not commonly provided or readily available.

As you read this work, please think about the stories and feedback you could provide to enhance this work. Here is an example of the type of stories wanted:

The Hammer Story

Leaders need to have a broader perspective on how to deal with people. To illustrate this need, I offer the following story which was told to me by one of our project superintendents, "Mike." Mike said, "I was a quality control manager on a project called the Marble Hills Nuclear Generating Station in Southern Indiana. The project superintendent, "Dave," was one of those hard-nosed superintendents who liked to throw his hard hat around and berate people. A rainy day brought Dave into my office to chat, and at some point during our conversation Dave said to me, "I just don't understand my guys. I think I have everything lined out and everyone understands the plan, then I go out there and they're not doing what I want them to do. So I go back out, pull everyone together and repeat to them what I want them to do. I yell and scream a little, throw a hard hat for show, and think I have everything squared away. But a few days later, I see they're STILL not doing it right, and that just drives me crazy!"

I could see that Dave was frustrated but I decided to take a bold approach. I asked, "If you hired a carpenter in the morning and by ten o'clock you realized that the only tool he could use with any success was a claw hammer, what would you do?" Without hesitation, Dave said firmly, "I would fire him!" I asked him why. Dave responded, "Well, if you can't use the tools of the trade like a pry bar, a screwdriver, and all the other tools a carpenter is supposed to use, then he's not a carpenter at all." Without missing a beat I responded, "So what makes you any different than that carpenter?" He gave me a blank stare, so I continued. "Seriously Dave, when it comes to management, the only tool you know how to use is a hammer. You're missing all the leadership tools that should be hanging on your superintendent tool belt. If you don't have those tools then you're nothing but a pusher, and pushers are a dime a dozen in this industry." Dave sat there looking out the window for awhile then got up and walked right out the office without saying another word. Three days went by before he walked back into my office and said, "Tell me that hammer story again!" We had a long discussion and he agreed that people don't always like to be pushed. Sometimes they want to be pulled, supported, trusted, or motivated. I realized," Dave said, "there are many other tools that are valuable in our tool belt when dealing with people.

Introduction

Cultural Creation

In Lessons of History, by Will and Ariel Durant, the authors attempt to summarize the themes they found in the eleven-volume work called *The Story of Civilization*. Anthropologists had long claimed that the destinies of empires were determined by war, morality, form of government, or a myriad of other things. In their book, Will and Ariel Durant suggest the rise and fall of an empire hangs in the balance of something called "cultural creation."

Roughly defined, a culture is comprised of the customs, arts, social institutions, and achievements of a particular nation, people, or other social group. The Durants argue that the customs of a culture, and specifically the behaviors that motivate those customs, determine a society's fate more than external conditions.

If internal culture plays a major role in a society's destiny, then it follows that the beliefs, decisions, and actions of a business will also decide its fate.

This book was written to shape the culture for all field builders. It is not a detailed step-by-step framework of what to do and how to manage your projects. It is a collection of stories, lessons, and encouragement that will hopefully shape the culture leadership of your company. This book was written specifically for Superintendents. Know that every word you hear was meant to be heard as if we're speaking directly to you. Think about what you're hearing and try to apply it to your project management style. If you allow these lessons to influence your way of thinking, you will see immediate results in your effectiveness as a builder.

Command with Purpose; Control the Outcome

Most of us have learned more from projects that have gone wrong than we have from projects that have gone right. In this industry, there are jobs that finish behind schedule, over budget, and with poor quality. There are projects that are filthy, out of control, and seem to have no specific leader. Interestingly—and unfortunately—these projects seem to be the rule and not the exception. This is unnecessary and is caused by contractors not having command of their projects and failing to control the outcomes.

Think about it like this: If an owner provides the correct funding for the project, and the owner or design-build partner provides an adequate design according to normal standards of care, then the remaining considerations are the responsibility of the builder. One might say external conditions such as weather, equipment, or economic conditions are out of their control, but with the right preparation, the builder will have planned for such externalities. The destiny of a project is in the hands of the builder, and as such, that destiny is fated by their attention to command and control.

There are many trade partners and vendors in our industry who can work at peak performance but are not allowed to. Take, for example, an outstanding electrician doing in-wall rough-in for the interiors of a building. She plans her work, coordinates the rough-in locations, prefabricates what she can, and has her manpower ready and geared to go for the benefit of her company and the project. What if the design then changes? What if other trades fail to coordinate with her rough-in locations? What if they fall behind and do not provide the work needed ahead of her? What if other trades leave open holes in the areas that are released to her? Even with all her preparation, the proficient electrician will lose production time, fall behind, perform re-work, and lose faith in the overall processes of the project.

In this case, the general contractor failed to provide suitable command and control. The general contractor owed the electrician a controlled project that would allow her to leverage her abilities. The coordination effort should have included all trades and its plan should have been enforced. The design changes should have been isolated so the main workforce was not derailed. The general contractor should have educated the other trade partners and enforced the concept of pulling work and finishing on time. And the general contractor should have created a system to see the potential problems that may have arisen in the work and prevented them through standard processes.

You can be sure when you see chaos on a project, schedules pushing behind, bad quality, and safety issues, that no one is in command and no one is in control. We must be in control of our projects to create remarkable experiences and shared success.

Part 1
Construction is War; Waste and Variation is the Enemy

General Patton declared, "The Nazis are the enemy. Wade into them, spill their blood or they will spill yours." We may agree with Patton's sentiment, but we are not army officers. We are, however, constantly at war with variation and waste. Patton's quote could easily be re-written for the builder to read as follows: "Waste and variation are the enemies. Wade into them, eliminate them or they will eliminate you." All-out warfare on anything that does not add value is the only approach that will allow us to reach our full potential. Every day is a battle against waste and variation, and believe me, we are at war. Everything in the construction process is literally trying to kill us on-site or waste countless hours of time and money.

Eliminate Waste and Variation

Waste and variation are intolerable. Sometimes it may be necessary to endure them during the pursuit of a more urgent goal, but if you must temporarily endure waste and variation, you should at least be sure that you never come to tolerate them. In fact, they should completely annoy you.

Imagine that someone scratched the paint on your new truck while it was in the parking lot. You worked hard to save up for that truck. You knew that eventually it would get banged up as you used it, but you'd hoped to be able to drive it for at least a few months in pristine condition. Now someone has carelessly opened their car door into the side of it and a chunk of paint is missing right in the middle of the dent they left. Not only did they damage your door, but the cowards didn't even have the guts to leave a note with their insurance information.

How do you feel as you imagine this scenario? Can you feel that absolute hatred for the situation? Can you capture that feeling in your mind? Good. Hold it. Let it constrict your blood vessels and turn your face red.

That is how you should feel about waste and variation. That is how you should feel when you walk onto project sites that are dirty. That is how you should feel when you open portable restrooms on-site and see no toilet paper. That is how you should feel when you see rebar on-site laying around with no apparent purpose.

Often, the reason we find ourselves battling waste and variation is due to a lack of discipline. At the beginning of World War II, the Allied forces had not yet fully proven themselves as a worthy opponent to the Nazi war machine. Sure, they had the drive, desire, and motivation, but there was still little to speak of as far as discipline and experience. In the battle of Tunisia, the Allied forces suffered incredible casualties in some of the first major battles of the war. The losses were devastating. When General Eisenhower ordered Patton into the fight there was not much by way of discipline, order, or guiding principles.

In the film *Patton*, George C. Scott portrayed Patton as saying, "Do you know why our boys got whooped in Tunisia? They have no discipline. They don't look like soldiers, they don't act like soldiers, and they don't train like soldiers." He was right. Patton immediately began instilling discipline into the troops and ordered standard military dress, training, salutes, and a slew of other disciplines that changed them from a rag-tag bunch of boys into fighting machines.

You must identify the enemy—waste and variation—and rest only when you gain victory against it. The fight should fuel you with passion and drive. You might think that sounds dramatic, but be assured that just as the lack of proper salutes among Patton's men seemed like a minor thing, tolerating waste and variation on our projects leads to major losses.

Waste

Waste is anything that does not add value. There are 8 recognized wastes in our industry: excess inventory, overproduction, wasted transportation, wasted motion, needless waiting, over processing, defects, and not using the combined skills of the team.

Consider inventory. Have you ever been on a project where all the fixtures or all the rebar were brought out only to sit, wait, get moved, damaged, re-ordered, moved again, installed with defects, sold to an owner who no longer has a choice, and then haunt you during punch walks with everyone who tours the building? Do you remember the last time you saw a stockpile of rebar get completely covered with mud on-site? Do you remember how it felt? It is all waste. The ignorance was waste as was the inventory, damage, and re-ordered materials. The lost productivity, added stress, missed football games at home because you were working late, lost profits, bad performance, hindered careers, all the difficult and contentious conversations, lowered morale, lack of pride, sleepless nights, and on, and on, and on is all waste, and we should not tolerate it.

Variation

Let's say you started making dinner for your family. You get all the groceries, start the burners, heat the oven, mix the ingredients, and get half-way through cooking the recipe. The kids now tell you they don't want that meal. Your spouse tells you he or she is working late and can't be home for dinner. You realize you are missing the eggs you need to make what the kids really do want, and in all the confusion you burned the food that was on schedule. How do you feel? What do you do?

Variation is any interruption to the flow of the project. It happens when information or plans change, when commitments are not met, and when consistency and flow are compromised. Many in our industry love creating variation for our folks on-site. Have you ever gotten word from the architect to make a change on-site, picked up the radio, called the foreman, and changed the plan? What happens next?

Think about a concrete crew on-site. They have forms mobilized, rebar on the way, layout performed, concrete and finishers scheduled, and everything is ready to go. Then you get a call on the radio saying, "hold off on the wall because there is a change." Or, "please stop work because we forgot an inspection." Or, "we need to put the bulkhead in a different location because we are missing information about a specific block out." Easy, right? Just stop. Just change. Just deal with it. We all know it isn't that simple. The formwork was needed elsewhere. The schedule is now behind. We just wasted all day mobilizing forms and must take them back. We have to cancel concrete. And, oh yeah, they can't reschedule for another two days, so we just lost another two days. More rebar needs to be ordered so the office is now interrupted. The suppliers are left fighting fires. No one now knows the new plan. Morale goes out the window. The engineers have to re-configure the layout. The work tomorrow needs to be rescheduled. The rest of the week needs to be planned again.

Variation is not easy, it is not fun, and it should not be tolerated. One seemingly minor change causes far more consequences than we realize all at the expense of our morale, personal lives, and all we care about. It might be necessary at times to please an owner or fix a mistake, but it should never be tolerated easily. It should make us cringe to create variation on-site. It should embarrass us. We should shield people from it whenever possible.

The Ongoing Battle

We are at war with waste and variation. They cause delays, ruin productivity, lead to confusion, create unsafe environments, destroy morale, and make you lose control. In all phases of the project, you should fight against waste and variation. There is no time, no circumstance, and no push so hard that we should tolerate them easily.

You may be put off by the tone of the last few sections. They may have seemed a little negative or may have struck you on a personal level. To reiterate: the criticism and condemnation is against waste and variation, not you. We should be against anything that does not bring you a positive experience on the job. You must expect more and be treated better.

Sometimes we get the idea that what happens in our industry is our fault or because of us. That is not so. You are a construction professional, and we want you to have professional expectations. Everything on-site, in the office, and as a part of safety, quality, schedule, cost, and your team should bring you joy in your work. If it does not, please do not settle.

The three pillars of lean are to learn the 8 wastes, take incremental steps to remove them, and to shoot before and after videos documenting your improvements. If something on your project is not remarkable--meaning remarkable for you, the workers, the owner, and your family at home--then expect more, leverage the team, and make a change. You are important enough to do that. That is why we should all fight against waste and variation together.

The Art of Attack

The Two Generals

Military generals can often be "pushers." When General Custer was at West Point, he constantly resisted authority and received 199 demerits in one year, just one demerit away from being held back. Custer pushed his men hard and is famously known for the "Custer Charge," a tactic that he employed ruthlessly throughout the Civil War. His unit had disproportionately high casualty rates, and his men paid a high price for their success.

The problem with this leadership style was shown in epic fashion when he pushed too hard at Little Bighorn. He rushed into a battle without planning or knowledge and with no supportive elements. His 208 men charged into the largest gathering of Plains Indians in history. Over 2,000 warriors met his assault as he was overpowered and was forced into a defensive posture known as Custer's Last Stand. The defeat was devastating. His foolhardy attack resulted in his death and the death of most of his troops.

A major problem with an aggressive charge is that it overcommits your manpower and resources in one specific direction without the required foreknowledge of possible outcomes. There is no guarantee the manpower and resources will provide the results necessary, and it may end in a devastating defeat. An aggressive charge forgets about the "art" and focuses only on the "attack."

Conversely, consider General George McClellan. He was an extremely skilled officer and West Point graduate thought to be a young Napoleon; however, he was hesitant and over-thought his tactics in the Civil War. He often second-guessed his plans and delayed acting which allowed the fast-moving and aggressive South to win several key victories. He botched the Battle of Antietam, which led to his removal from command and may have been the reason why the war continued for another two-and-a-half years.

Supers and Generals

The *art of attack* requires you to balance the total workflow of the project. If you evaluate the balance between planning and execution you will quickly see that one unit of planning is worth multiple units of execution. One hour of careful planning can easily save hundreds of man-hours, prevent needless material waste, avoid scheduling conflicts, and increase jobsite morale. There are times when a brute-force approach becomes necessary, but brute force is much more efficient when applied strategically by first planning where to apply that force.

Planning also allows you to know where you can advance responsibly. If a Custer-style charge is part of the plan, then so be it. It just needs to be preceded with a responsible amount of quick planning.

We must also be wary of over-planning and making excuses to delay work. Just as McClellan's hesitancy to act led to devastating defeats for the North, delays on worksites caused by over-planning can cost the company money and waste our employees' time.

Sun Tzu said in *The Art of War* that "though we have heard of stupid haste in war, cleverness has never been seen associated with long delays." We must not be hasty. We cannot be stupid, ignorant, and unlearned in the art of planning and organization to the expense of our men and the cause we are pursuing. But we also cannot sustain long delays. Waiting, when unnecessary, brings with it many unintended consequences that are not always easily seen but are heavy prices to pay.

The *art of attack* requires finesse while maintaining an offensive posture throughout the project at the right times and places, not thoughtless pushing forward or fearfully delaying work. Both result in waste.

The Two Supers – A Story from the Field

"I am NOT canceling the concrete placement! Once I do that, the guys will think I'll cancel it every time!" This is what one superintendent passionately told his team when asked what to do about a placement that was falling behind.

He and I were two of three superintendents on an eighty-million-dollar commercial building project. This superintendent was a "pusher" and production was all that mattered to him. He would drive his crews almost to exhaustion while working overtime on Saturdays and holidays. This came at the expense of quality, safety, organization, and cleanliness. Deck placements were often missing embeds, columns were placed 2-1/2" out of skew, and walls were missing reinforcing because placements were made in haste. His intemperate pushing came at the expense of his people, product quality, and job morale. Though this superintendent was always charging in and pushing his workers, he never understood the "art" in the *art of attack*.

The other superintendent I worked with was the opposite. He was all "art" and no "attack." He was analytical and hesitant. In his mind, everything had to be perfect before he began his projects. He was constantly late and passive with scheduling trade partners. I always noticed he was toiling away, perfecting nonessential parts of the project. On one particular part of a project, I needed to know how my scopes and schedules fit in with his, so I put his plan together for him. His indecisive over-thinking not only put his scopes behind but eventually caused the entire project to fall behind by three months. The project superintendent eventually assigned half of his scope to me; after all, I had planned it for him.

Clearly, these two superintendents were diametric opposites. They never understood the *art of attack*.

I was in charge of the largest scope of the project. I completed my scopes of work on time, ahead of schedule, and in support of all other scopes for the building. The intricacies of my scope were orchestrated to work around the other scopes with little impact. The systems installation was masterminded to begin with an early start even though it took some creative engineering. This allowed building systems to be up and running on time. The team was never waiting on my scopes to be completed. This was the result of preparation and strategy, not hasty pushing. It came from thinking, not working people harder. It came from decisive action. It came from the *art of attack*, which is a strategic and balanced approach to advancing to victory.

Time of Exposure

General George S. Patton warned against entropy when he declared that "every soldier should realize that casualties in battle are the result of two factors: first, effective enemy fire, and second, the time during which the soldier is exposed to that fire. The enemy's effectiveness in fire is reduced by your fire or by night attacks. The time you are exposed is reduced by the rapidity of your advance."

Every builder should realize that casualties to our success are the result of two factors. First, the effect of waste and variation on plans, and second, the amount of time we allow to be lost to waste and variation. The negative effects of entropy can only be attacked by a robust and effective effort to organize and overcome. Exposure to waste and variation is reduced by the rapidity of your advance.

Something will always go wrong, and the longer we wait, the more chances there are to allow that to happen. Every day we wait, materials go missing or become damaged. Every day we wait, people become busier with other pressing matters and disengage with the needs of the project. Every day we wait, we are missing the opportunity to fix problems that surely exist elsewhere.

Sun Tzu advised, "Rapidity is the essence of war: take advantage of the enemy's unreadiness, make your way by unexpected routes, and attack unguarded posts." As builders, we take unexpected routes by strategically leading a team to completion. Take advantage of the enemy's unreadiness by solving problems before they're allowed to alter your destiny. Never allow the enemy—waste and variation—to grab hold. Take decisive action and reduce your team's exposure to enemy fire.

Setting Air Handlers - A Story from the Field

I recall a large project where we were trying to start up the air handlers to air condition the building with cool air. This was critical to the success of installing finishes. It was August, and the projected air-on dates were anticipated the following June. I began drawing the plans for these expensive machines a total of ten months in advance.

I was criticized for planning that early, but I knew things would go wrong. I knew roadblocks would try to hinder our progress. I knew enemy fire would be upon us and that we needed to advance where we could. Therefore, the massive chilled water piping installation was scheduled early at my request. The electrical startup date was scheduled five weeks early in anticipation of delays. The water tie-ins were moved up. Everything was moving consistently toward the target dates in a slightly faster, but responsible pace.

All this planning paid off. The chilled water lines ended up being delayed due to material fabrication and were completed only two weeks before we needed that critical 42-degree water. The permanent power required an additional month to energize due to delays with the bureaucratic, government-sanctioned power company. The piping took an additional two weeks to flush before systems could be operational.

Did all of that slow us down? Did it prevent us from hitting our target date? Absolutely not! Why? Because contingencies were built in and the driving pace with which we built the separate systems came together at the right time. We did not wait. We did not second guess ourselves. We did not assume the best-case scenario would happen. We attacked. We reduced the time we were exposed to enemy fire by increasing the rapidity of our advance.

Avoid the Defensive Position

Defense Is Stagnation

There may be times when an army, nation, team, or individual needs to remain in a defensive posture, but it is not an effective way to lead. Granted, if the options are ruin or a defensive posture, choose the defensive posture, but regroup to attack once circumstances allow. There is no safety or victory in staying on the defensive. As Patton observed, "Nobody ever defended anything successfully, there is only attack and attack and attack some more."

Historically, defenses eventually fall to ruin. Only in attack do we keep our enemy on the run and preserve our forces. Only on the move do we stay safe. When stagnating or on the defense, we become the object of attack. We make no progress, and our fall is inevitable. If we must be in a defensive position, we need to use that time to redirect and plan a counterattack.

Once a group is on the defensive, they are handicapped. Consider the Nazis in World War II. Their Blitzkrieg tactics of attack enabled them to quickly conquer most of Europe. The tactics were so effective that France was conquered in six weeks and Denmark was conquered in a mere six hours, with a handful of other countries falling somewhere in between.

However, once they found themselves as occupiers, the Nazis had to change their behavior from one of aggressive offense to that of defense. They had to defend 1,600 miles of coastline from the impending Allied invasion. The prospect of defending such a large area brought huge problems--everything from not having the steel and concrete needed to build defenses to not having the manpower to garrison them. When Patton had them on the defensive, they had to help stragglers, destroy equipment and ammunition, obtain mobile supplies, re-group, and try to rally time and again. So, not only did they need to carry on the normal necessities of war, but they also had to re-establish the critical momentum Patton's army already had.

Behind the Eight Ball

When we sink into a defensive position on our project sites, we find ourselves behind the eight ball. In the game of eight ball, the eight ball is the last ball that a player sinks into a pocket. Being behind the eight ball refers to times when the cue ball is positioned behind the eight ball and cannot get a direct shot at the other balls. In this situation, there is no good move available.

Failure to plan will put the team behind the eight ball and keep them from advancing on the project. Consider a project with no advanced safety planning or implementation. Once someone gets hurt, the project is unmanageable. The team does not have any time to advance or overcome. They can only react to circumstance. Likewise, when a team has allowed a project to become unclean and disorganized, they cannot instantly implement a habit of cleanliness and organization because bad habits have already been formed. With all the time spent going back and cleaning up after themselves, they do not have time to prevent future infractions.

Firefighting

Being behind the eight ball puts the team in firefighting mode. Everything is an emergency; everyone's job becomes putting out fires. A firefighter at work is not trying to avoid water damage. He isn't hesitant about breaking down a door. He isn't focused on anything other than putting out the fire. When he enters a burning building, he hacks through the roof, drenches the home, and damages anything in his way to extinguish the flames. And in some cases, it is deemed safer to just let the house burn down.

The same is true in life and work. When superintendents fall into firefighting mode, there is no room for preservation or prevention. When waste and variation are destroying our future work, the clock is ticking. We cannot sit down to strategize and our knee-jerk attempts to solve problems often end up causing more damage to the project.

Losing

Not winning or advancing demoralizes a team and renders them ineffective. If the defensive position is not enough of a disadvantage, the demoralization of your team will eventually tear it apart. Humans need to have a win from time to time to feel effective. No pep talk, sermon, or rally can build the team up the way a tangible win can.

Consider the story of Constantinople and its eventual fall. In 324 AD the Roman Emperor, Constantine, chose the existing city of Byzantium as the new capital of the Roman Empire. Rome was no longer fit to be the capitol. He chose this city because its position offered easy access to trade routes and it was closer to the frontier of the empire, but mostly because it was so easily defended. He named it Constantinople and built defenses that remained impenetrable for over 1,000 years. The residents of the city relied on their location and fortifications as a defense and over time became complacent.

After centuries as one of the finest cities in the world, Constantinople became the target of on-going attacks from neighboring civilizations. Eventually, weakened by political and religious in-fighting, economic downturns, and the plague, the great city finally fell in 1453 to the Sultan Mehmed II, leader of the Ottoman Empire.

This pattern has replayed itself over and over throughout history. When cultures, companies, or people aren't actively and regularly winning, they are on their way to losing, even if it takes some time.

Becoming Distracted

The primary focus of any leader should be strategizing. Battles are often won or lost before the first gun fires, in large part due to effective planning. When leaders are forced to focus on putting out fires, they are taken from their primary role of planning the next improvement. Once the team is in emergency mode without a clear plan, the unit is leaderless. Once reactive behavior takes over, there is no control. This is true for leaders in any position.

Part 2
Prepare Your Foundation

As builders, you know that you must lay a proper foundation for your project to be successful. Likewise, you need a good foundation of beliefs to be successful leaders. The following ten principles, if held as rock-solid beliefs, will ensure that you can move forward in an effective and successful way.

No one would go into war without following tried and true military principles of war. Attacking our projects with the following principles is synonymous to attacking in war with training, helmets, equipment, and a proper chain of command and control. We can then attack in a steady, clean, planned, and controlled way without deviation. If we run our worksites without these principles in play, we have begun with haste, not artful attack. The following principles must be understood and followed in any circumstance on-site for any project.

Principle 1 – Prioritize Cleanliness, Physical Organization, and Minimal Inventory

Cleanliness, organization, and minimal inventory are the best indicators of a project's health and stability. If a project is not clean and organized, it's not going well. Why would an unclean project be considered acceptable? A clean job is a safe job. A clean job is a job managed well.

If someone was lying on the ground at work, you would assume he was either sleeping on the job, hurt, or dead. You would investigate immediately. You'd try to wake him up then check his breathing or his pulse—key indicators of whether something is wrong.

Cleanliness, organization, and inventory levels do that for a project. All you must do to check the pulse on a project is to walk around. If it's not clean, it is in trouble, and it is not creating a remarkable experience for you. On a dirty project, you can't assess safety, quality, morale, materials, flow, excellence, schedule, or anything else; therefore, none of those things are being properly managed. If you want excellence, you will focus on cleanliness. Beware of people who do not understand the importance of cleanliness because they are displaying their mediocrity.

There is never a situation that should not start with proper cleanliness, organization, and right-sized inventory levels. When it is not clean, we are at risk for trips, falls, and falling objects striking other workers. When it is not clean, we are not delivering on customer satisfaction. When it is not clean, workers are subconsciously being shown the standards of care they should take in their work. When it is not clean, we tell workers how they should treat us and the bathrooms on-site. When it is not clean, people are not productive. A good worker is a clean worker. A good crew is a clean crew. A good foreman is a clean foreman. A good GC team is a clean GC team. That's it. If it is not clean, we have dysfunctional leadership in one form or fashion, and we do not have operational control.

This is not to say it is easy. That is precisely the point. It is not. It's hard. That is why so many justify it. They do not have control and they are embarrassed because of it. Then, they justify. We should expect more, and you deserve better.

A clean site is a stable site, and stable sites and systems are the only environments where improvements and respect can flourish. So how do we do it?

We must expect more. First, decide to do it. We need tenacity and high standards. We always say, if the leaders care about it, then it will get done. If you see the leader care, the rest of the workers on the project will care. If you see the lead superintendent picking up trash on-site, you will see others pick up trash on-site. So, decide, act accordingly, and that's most of the battle won.

Once you have decided, follow through and do not give yourself an out. Leave no room for retreat. Draw the line in the sand and let no one pass without consequences.

The trick for this principle to work is to seek perfection. Perfection must be the standard, and the setpoint of the project and the superintendent must be set to perfect. A setpoint works just like a thermostat. If your thermostat is set to 74 degrees and you leave the door open for a while, the room may reduce in temperature for a moment, but it will return to 74 degrees when the door is shut. Additionally, if you cook in the oven and increase the room temperature to 80 degrees, it will quickly return to 74 when the oven is turned off.

The same is true for setpoints on our projects. If the superintendent's setpoint is set to dirty, then it will always return to that state. It may be cleaned temporarily but it will get dirty again, and if it gets really dirty, the superintendent may order it to be cleaned a little. But it will always return to just good enough.

The setpoint must be set to perfection. If it is, the state of the site will always conform to a higher standard.

Excess inventory carries with it some of the same issues as uncleanliness and disorganization. With too much inventory on-site, things are in the way. This keeps you from being able to see what you need to see. It reduces the flow of work and causes project teams to become part-time moving companies instead of project workers. Materials should arrive at the project and go in place just before they are needed. There is little merit to bringing too much inventory to the site because it slows down production and uses up a valuable commodity—on-site space.

Modeling Expectations – A Story from the Field

You may be thinking that this is all blue-sky dreaming, but when I began building the research laboratory in Tucson Arizona, I was determined to have a clean project. I decided we would never have a composite cleanup crew. We would never have temporary laborers on-site. And I meant it.

If there was a mess on-site, I politely asked the crews to stop all work and clean it up. To show the level of cleanliness I expected, I led by example, helping to pull items that did not belong into hallways or open spaces for removal so the space could be swept. The action sent a clear message and I would bet we had a job as close to perfection as you could ask for in the state of Arizona. Everyone commented on it. It was so satisfying. After a few weeks, it began to be self-sufficient too. Why? Because people's standards were higher. Their tolerance for messes decreased and they became addicted to a clean site.

I remember leaving for two weeks on PTO. Was the site still clean when I got back? You bet it was! They held one another accountable, and it was great. All of us as team members would help, remind, pick up trash, and pitch in. It became a project motto and rallying cry: "Clean and Steady," we would say. And we said it in English and Spanish so everyone knew it. I believe it worked because we explained in daily huddles why we did it. It worked because workers on-site are not animals and they don't like being treated that way. It worked because we set the example. It worked because we cared.

Principle 2 -
Steadiness and Flow

There is a popular game used to teach Lean Construction Management called Parade of Trades. Seven participants sit together around a table. Each one represents a contractor for a different scope--concrete, steel, façade, and so on. There are 35 chips on the table that represent 35 pieces of work that must be completed to finish the project. The goal is to pass all these chips through to the last contractor as quickly as possible to simulate the completion of the work on the project. To complete their work, players roll a die and pass that number of chips to the person next to them. If someone rolls a six but only has one chip at their station, they can only send one chip forward. If they have six chips and roll a one, they can only move one chip. When everybody has rolled, that represents one week of work, and the process is repeated for as many weeks necessary to move all the chips to the end. Often, this game is played with different teams competing against each other to be the team to finish their work in the fewest number of weeks.

Let's say that at one event, the red team got all 35 chips to the end in 26 turns. When you tally the number of men on-site, you get 380. The maximum inventory at any one location was 10. So, in 26 weeks, the red team moved 35 units of work down the line with 380 people and an inventory level of 10 units per week. Pretty good.

How did the blue team do? They completed the work in 21 weeks with 280 people and a maximum inventory buildup of just five in a given week. That means the blue team was five weeks ahead of schedule with 100 fewer people on-site and half the material on-site. How did they manage that?

The tricky secret is that the red team had a regular six-sided die with numbers ranging from one to six. The blue team had a die that could only roll fours, threes, and twos. A week with the normal die would be something like 6-5-5-2-2-1-1 and would represent an attempt to move quickly at first followed by a slow down because of an interruption and eventually ending up stuck.

A week with the blue team's die might look like 4-4-3-4-3-2-2. This is synonymous with the concept of maintaining the flow of work. A flow has very little variation. When that variation is eliminated, the chips (or the work), could flow from one end to another without getting held up by the overwhelming rush from rolling a six or the painful crawl of a one.

What is the lesson here? Before making that explicit, let's reflect on the current state of our industry.

We often hear people say "We need to get out of the ground fast" when we can influence a fewer number of contractors. This is correct. We also hear that we need to be aggressive with complex and unknown areas and scopes on our projects. This is also correct. Sometimes though, well-meaning folks will apply both of those concepts to the entire project and say things like "Advance the schedule whenever possible," or "I want all my materials here now," or "Just bring it," or "I am a pusher," or "Keep pushing everything you can for the schedule". These people are trying to roll sixes in Parade of Trades which will later cause a mess of ones.

Consider this: What if we told a superintendent to slow down a little and keep a steady flow and an even pace? What would he say? He might say we know absolutely nothing about construction and should go get another job. And yet, the data shows he will finish in 26 weeks with 380 people on-site with an inventory of 10, like the red team in Parade of Trades. This happens because people think movement equals production, which is not the case. All this movement is actually waste. People start pushing and creating variation because it makes them feel good. It gives an impression of progress. But it requires twice as much material, a hundred more people, and five more units of inventory in a week.

What happens when a superintendent won't keep a flow in the schedule and hold dates? If you are a trade partner, how would you react? You would possibly keep more people on-site and most definitely keep more materials on-site.

What happens when materials pile up on-site? You guessed it--production slows down. Everything slows down because part of our workforce is dedicated to managing, moving, inventorying, fixing, replacing, re-ordering, and organizing materials.

Here is an example: Imagine a factory with an assembly line that produces a certain number of finished items every hour. The throughput is the rate at which the factory can process the raw materials into finished items. Say you have four machines in the factory that work together to produce the final finished project. Items need to be moved from one machine to the next for that machine to do its part in the process. The first machine can work on four items per hour, the second at two items per hour, and both the third and fourth machines produce four items per hour. What is the throughput of the system per hour?

Did you say two? This may surprise you, but it is likely going to be 1.25 or 1.5 finished items per hour. Consider what happens between the first and second machines: an inventory of materials begins to pile up. Manpower is then allocated to manage the inventory. Space in the factory diminishes. People who would otherwise be running machines are now managing the machines and the inventory. Workers down the line are waiting, and more resources are needed to manage materials on the third and fourth machines. Waste increases and the speed of the system decreases. Therefore, the throughput is 1.5 or fewer finished items per hour, instead of two.

When anyone on-site says, "Work everywhere at full efficiency," or "Keep pushing," or "Bring all the materials here now," or "Don't slow down anything," or "I want workers working everywhere on-site with no empty areas," all they are doing is slowing down the speed of production by increasing inventory and creating a lot of wasted jobs for people who would otherwise not be needed on-site. Why? Because we needed flow, and what we got was variation. If this was the Parade of Trades, we needed to roll consistent threes and fours instead of sixes and ones.

The answer is not to push all the time nor is it to flow all the time. The ideal is somewhere in between. Anything on-site that can be made to flow, should. If we are coming out of the ground or have a complex area on-site with high-risk unknowns, those may be good reasons to push. The point is that pushing comes at a risk, and flow will always reduce materials, manpower, mistakes, and the time it takes to do something. So, if you can create flow, do it. If you must push an area, know the consequences and do it only if you must put work in place early to vet mistakes early. But for no reason should you push everything on-site and create variation. Don't roll sixes and ones when you can roll threes. The *art of attack* means artfully taking advantage of opportunities to plan, prepare, and move strategically.

If you hear someone ask to create variation in schedules and flow, be skeptical. If you hear folks ask for an increase in inventory, give it a second look. We want to create stable environments, keep workers installing the work they planned for the day or week, and have a plan for everything.

Principle 3 – Value Planning to Create Stable Environments

There is a commonly offered piece of advice to "plan it first, build it right, and finish as you go." Everything we do must have a plan. We do not go into battle until we have a plan. In fact, we should not go into battle until we have a plan A, B, and C. It is better to even have a plan A, B, C, D, and E.

In *The 33 Strategies of War*, there is an entire section dedicated to planning. The author, Robert Greene, details how Napoleon would lay down on large maps that spanned from wall to wall on the floor in large rooms. He would study the terrain, the roads, the cities, and the strongholds of his enemies. He would go through all the scenarios in his mind and plan what he would do next. He would do this for days at a time.

During his campaigns, he would have setbacks, delays, and unexpected events, but he would always re-group and advance. He always had a plan B. He knew no plan would endure engagement with the enemy. He knew things would change. And he had options to keep advancing and keep winning. Artful attacking is about planning and taking advantage of opportunities in a steady, intentional, and organized way.

Patton famously said, "A good plan violently executed now is better than a perfect plan executed next week." This rallying cry still indicates the need for planning. He did not say, "Attacking without a plan is better than a perfect plan next week." He said "a good plan." And a good plan is one that has been researched to anticipate problems that might arise.

We should have an A-plan, along with plans B, C, D, and E waiting in the wings to deal with problems we can anticipate. We hope the old days of supers running around pointing, directing, yelling, re-directing, and fighting fires are over. The team should not only follow the leader, they should follow the plan. When leaders try to be a modern-day Moses on projects and do everything themselves with the plan in their head, they are leading their team to ruin, and they need to make a change, just like Moses did.

When you hear the word *attack*, I want you to remove the mental image of people running into battle. Change that mental image to a scene of soldiers working together to execute a planned assault under the direction of a leader who has explained the plan to them ahead of time. We have to direct field crews the same way.

Imagine yourself on your jobsite. At 2 pm, you get a call from the designers who order a change to something that is in progress on-site. Our first reaction would be to just call the foremen on the radio and tell them to stop or change, but that should not be the first response. We should consider first asking if we can let the crews finish out their day as planned and regroup before the next day. If the change to the work order is more urgent than that, we need to gather all available information, ensure that our preparation is complete, make sure we have financial approval, and print documents showing the new change. To put it simply, we must develop a new plan. Then we can walk outside, ask the workers and foremen to huddle up, review the new material, and give them the time they need to process and re-group.

In no case should we have a knee-jerk reaction and tell people to just change and switch tasks. We need to take the time to develop the pre-task plan, acquire the right tools, and make sure everyone knows what to build. If we do not go to the workers with this information and give them time to change, we are mindlessly sending them into battle with no strategy. It is not right and should not be done.

Great leaders try to lock in the plan for next week. If they cannot do that, they should at least lock in the next day. If they cannot do that, we owe our workers an apology and a huddle to regroup. We only send people into battle, or into work, with a plan. Period.

The ideal job site is an environment where workers return home and report that they were bored at work. No fires, no accidents, no excitement, no treasure hunts, no drama, no unclean restrooms to complain about, nothing. Just stable, predictable, quality work, day in and day out. There should have been a preconstruction meeting, a first-in-place inspection, and a material inspection. Workers should have lift drawings for their work, on-time deliveries, tools and equipment, space, training, and time. They need proper bathrooms, a decent lunchroom, an effective communication system, and anything else they need to put work in place.

When do we make money as a company? When a worker is putting work in place. When do you, as a superintendent, make the company money? When a worker is putting work in place. When does a director make the company money? When a worker puts work in place. Our industry has focused for so long on trying to make project management systems more efficient when we should have been trying to make the workers' environment on-site more efficient and stable through planning and operational control. Our focus has been off. We need to get it back where it belongs—on the worker on-site. If we do not properly plan so we can provide a stable environment, we have work to do.

Stable environments are hard to create but must be our focus. We need to understand that mental capacity and mental discipline is an exhaustible resource. There are only so many good decisions a worker can make before he or she becomes mentally exhausted, and when a worker becomes exhausted, they are unsafe. On a chaotic site, a worker has to watch where they step, look up, step around trash and materials, watch out for the forklift, try to find the hammer, go look for materials, read three sets of drawings to figure out how to install something, remember to wear safety glasses, watch out for the leading edge, answer calls and texts from home, etc. There is no way that workers can focus and be effective when working amid chaos. Their discipline and mental focus are quickly exhausted.

We want our team to show up for work, get clear instructions, pull their tools from a clean gang box, walk across a perfectly clean floor, walk in clearly delineated pathways, install per the lift drawing, be at lunch according to the posted rotation, and so on. If most of the site is fixed with stability and routines, the mind can focus its attention on being safe and on the work. That is why stable environments are so important. Waste cannot be allowed to drain precious mental resources from the focus of working safely on-site.

Nothing can be improved in chaos. Can you raise the bar without having set the bar? Can you increase the quality of a minimum standard without a minimum standard? The answer is no. Only in stable environments can anything be improved. Only when a minimum standard is in place and followed can the team take it to the next level. You cannot improve anything without stability. I often hear folks ask, "Why would we advocate for more meetings on-site? Aren't we trying to be lean? Aren't we trying to be more efficient?" By holding meetings on-site, we are conveying the plan and making sure everyone has bought into it in an effort to create a stable environment. We do that by taking ourselves out of our comfort zone so that the workers can settle into their comfort zone. We deal with a little more sacrifice and effort, so they can do what they do best.

Principle 4 - Everything Good Thrives in Accountability

Have you ever been in a situation where you find workers behaving unsafely or unprofessionally? When you step in and talk to them, you often find that they did not know what the expectations were. Surely, we have all experienced this. If they are telling the truth--and we give them the benefit of the doubt--then they cannot be held accountable for something they did not know. But, if the superintendent has communicated with them at their level of understanding, then they can be held to a proper level of accountability.

Everything good thrives in accountability, openness, and teamwork. Patrick Lencioni wrote several books about team building and management. In *Five Dysfunctions of a Team*, he presents a formula for building and maintaining a team. The steps he lays out are building trust, engaging in healthy conflict, setting goals together, holding each other accountable, and performing as a team. The way we can judge if we are living up to this model is by assessing our workers' performance. A team will not perform if they do not have accountability based on team goals that are formed in trust and grounded on everyone weighing in and buying in.

Do you hold your trades accountable? If the answer is no, then you have room to improve as a leader. Do you hold your team accountable to perform according to team goals? If not, it is likely you do not trust each other. Performance equals accountability to goals the team sets together based on buy-in and trust. Thankfully, it is that simple.

We all get too hung up on thinking someone is nice or mean or whether someone has the right to tell us something based on whether they are our boss or not. We decide whether we should listen to people based on these assumptions and judgments.

This is not a good way to look at things. There is no nice or mean when it comes to accountability. There are such things as nice or mean in the way we deliver a message, and obviously, we should always deliver instruction and feedback in a safe and healthy manner. But the truth is that accountability is respect; it is kindness. It just needs to be conveyed in a respectful manner.

As a leader, you need to consider how to convey information about expectations and safety in a way that shows respect and appreciation for your team. It may take practice, but we are all accountable to each other as a part of the team and we should be modeling respectful behavior for our team. All members of our team have the right to say something in a respectful manner. There should be no worries about who is the boss or not the boss or whether what we are saying will be viewed as being nice or not nice. When it comes to accountability for safety and performance, everyone should feel empowered to speak up. Hopefully, even new employees feel that they have the right to call out the directors on a team if they see a reason. If this is the culture, the concept of a team is very powerful.

This concept might be hard for some superintendents to swallow, but you are already a member of a team and therefore accountable to a team. Consider extending the same level of respect and accountability expected of you from your management team to your team of workers.

There will be many good people on your team who will not feel comfortable calling someone out. There are many who do not like conflict. What do you do in that case? It takes practice. If you see someone who needs to be corrected, speak with that person directly and respectfully about the proper expectation for their behavior. If someone comes to you with a team member's failing, encourage a face-to-face interaction between the two parties. As a supervisor, you will act, not as "muscle" to back up only one side of what happened, but as moderator. Thank the one team member for bringing the issue up, ask the other to behave differently in the future, and then thank them for listening. If you observe a team member knowingly enable another team member to act dangerously, call them out for not speaking up and ask them to tell the harmful person how they should act. You must have the courage to practice difficult communications and reinforce your team members' practice by showing that you approve of workers respectfully correcting each other.

We must care enough to do what is right and leverage the power of the team to do it. You cannot watch every member of your team at all times. Our workers need to feel empowered to identify behavior that is dangerous, wasteful, or unprofessional because they know that they succeed or fail as a team. Delay of work due to an injury affects them as much or more than it affects the company. They need to be as invested in preventing bad behavior and waste as you are. No one on the jobsite should tolerate things that can lead to harm.

Consider the idea that "the culture of any organization is shaped by the worst behavior the leader is willing to tolerate." Does that sting? Does it ring true? If so, it's time to raise your expectations.

Failing to cultivate team accountability for individuals following safety guidelines is a form of disrespect, plain and simple.

Principle 5 - Communicate for Understanding

Communication is key for accountability, yet it is greatly misunderstood in the construction industry. Safety meetings, trade partner meetings, OAC meetings, and orientations have mostly become check-the-box, I-told-them type of meetings. Telling, telling, telling, and more telling. All telling and no understanding.

It might make people feel better to hold those meetings, but they serve little purpose if the people in those meetings only take in 2-6% of the information presented. Consider the number of hours that wastes in a week. Let's say that in the course of a week there are two orientations, one safety meeting, one trade partner meeting, one OAC, and two hours of other mass-meeting interactions. That totals approximately seven hours of meetings with telling as the primary focus. Seven times 60 minutes is 420 minutes. Six percent of 420 minutes is 25.2 minutes. That is less than a half hour of effective communication from seven hours of meeting time. Do you know anyone who would do seven hours of work for a half hour of return?

Why do we accept that much wasted time in our project meetings? Perhaps because we are told to have the meetings but haven't learned how to run them effectively. Or, we are confident in running a meeting but haven't yet learned how to communicate for understanding.

Here are the measurements for communication: Do the workers understand what is being said? Did they pay attention? Do they speak the language in which the information was given?

Ask yourself--How much of the plan do the workers in the field understand about your current project? How much do the foremen understand? How much of the safety orientation does every worker know and remember? How many of the safety topics do they remember? Do they remember the project rules?

If you frankly consider the answers to these questions, you will likely find your workforce is far from perfect in these areas. Few people read or understand our P6 schedules, fewer folks remember anything from orientation, and no one can visualize a complex list. Most workers need to see things visually, with fewer words, more frequently, and in their native language. Communication should be for understanding, not telling.

Consider honestly how much of the daily pre-task plan the workers understand and remember during work. The answer is very little. No one, even the best of us, would read that much text and remember it.

In our industry, most people in the trades are there because they did not like book learning. This is not meant to be degrading. Everyone has different skills and preferences. Thank God for them. But knowing that, why would we expect them to read through and retain a long pre-task plan every day? Why are our plans not visual? Why are they not simpler? Why do we not focus more on having translations available? Why do we not push people to interact in meetings?

Every worker should come on-site every day and know what is going on, where to go, what materials to use, how to build it, and why it is important. And they should understand all the instructions they've been told.

We have an obligation to get information and materials to our people. If that means we have to draw a picture to make the instructions easier to understand, then we do it. If we get creative, we can find ways to communicate effectively. Some suggestions are to test people after orientations, make visual schedules, draw the safety plan for the day, draw installation instructions, tell stories, and provide mockups. The challenge we all have is to get all communication to the point that it is understood by everyone. Easy? No! Possible? Mostly! Worth the effort? Absolutely! We must communicate for understanding in all we do.

Principle 6 - Hold Respect for People as Your Guiding Principle

What should we remember on every project? Respect for people! That's why we do everything. We take care of the customer because we respect them, their staff, and their end users. We treat trades well by ensuring our worksites are clean and materials are in place when they are needed because we respect our trades' productivity. We provide adequate facilities, bathrooms, lunchrooms, and treat employees fairly because we respect them. We do not tolerate safety violations because we respect people's lives and the well-being of their families.

Who or what do you respect most in your life? Some people might answer that they respect a few specific individuals. Some people respect money. Some respect their own careers. And some respect their families. There is nothing wrong with any of these approaches. There is, however, an answer that ultimately encompasses all these things, and that is to respect humanity—to respect people. Everyone! This attitude is a game-changer at the workplace because it puts every aspect of our work into focus. There is no philosophical question about anything on-site that cannot be successfully answered with the concept of respect for people.

Ask yourself the following questions: Why do we advocate having nice bathrooms on-site? Why do we encourage open and collaborative communication? Why do we sometimes put out misters where workers are waiting to haul materials in the Arizona summers? Why do we host barbecues? Why do we keep things clean? Why do we try to make sure all contractors on-site make money? Why do we not tolerate bad behavior?

Respect for people should be the priority that motivates us. If someone does not understand that, they may be confused about why we are so nice about maintaining our bathrooms but appear mean when it comes to an intolerance for bad behavior. We have nice bathrooms because we respect people, and we do not tolerate unsafe behavior because we respect people.

Additionally, any effort you make as a leader will be more difficult if you are not motivated by respect. If you are punishing people to punish them, then you will succeed in building resentment rather than bringing about the needed change. If you administer consequences to someone with the goal of keeping people safe, then your manner will reflect that, and you will inspire effective change.

How do we demonstrate respect for people? We respect people when we keep the jobsite perfectly clean and organized. We respect people when we remove waste and wasted effort. We respect people when everyone is safe and behaves professionally. We respect people when they are able to care for their families in a balanced and blended way. We respect people when we protect their finances. And we respect people when we communicate. If we do not do these things, we have work to do.

If we create positive experiences for our folks on-site, that helps us create remarkable experiences for our primary customers as well as the neighbors of the project site, the workers' families, our trade partners, and the end users. We have an obligation to take care of all these groups of people who are affected by our organization. There is little success in taking care of one of these groups at the expense of others.

This kind of thinking is challenging and requires us to be at the top of our gameplay. We are building people who build things for people. It is not about personal control, power, or title. Everything we do is to elevate and respect those around us. This giver mentality will change your life and career.

Principle 7 - Be a Team Player

The builder has an obligation to create positive experiences for people. The primary way to do this is to be a great team player. Unfortunately, the position of a builder and leader on the project can be a stressful and overwhelming position. We cannot let that push us into an unhealthy mental state that causes us to treat others badly.

General George S. Patton is known for many things, including his military success, but he's also known for the fact that he was severely reprimanded for slapping two soldiers he thought were cowards. Yes, he was arguably the most skilled military general in American history, but his temperament may have kept him from reaching his full potential as a leader. If it was not for his bad behavior and inability to act as a team player in certain situations, he would undoubtedly have been even more successful in combat against the Nazis. Instead, he was put in the proverbial doghouse for extended periods of time to humble and correct him.

How many lives were lost because of his inability to control himself? How many lost opportunities were there because he had to be in time out? We may never know. What we do know is that he was passed up for promotions and assignments because of it. We also know General Eisenhower would not relieve him because he was too valuable to the cause and considered indispensable.

Many builders in our industry, and especially superintendents, find themselves behaving similarly. They act out, yell, demean, say insulting and harassing things, limit their potential yet are still considered too indispensable to let go. This is a very unfortunate situation that causes frustrating delays to the builder's career advancement and collateral damage to the project. This is all unnecessary.

It is detrimental to the project to have inappropriate behavior from the star players. Project leaders must be able to direct the project in a respectful, professional way. Some superintendents seem to believe that running a site is like running a military campaign and they develop a leadership style akin to Patton's, relying on the force of their temper to intimidate others into action.

Is it possible to "go to war" at work and still be effective leaders? Only with a change in attitude.

There is no need to introduce unhealthy conflict to combat waste and variation. Likewise, your team may be meeting their goals out of fear of your wrath, but they will perform even better if you have shown that you properly maintain the flow of materials and work, are clear with your communication, and uphold your high expectations. You can gather your team, drive forward, administer consequences, and attack the task, all without fighting.

If you find yourself scoffing at this idea, it's probably because you have room to improve. First, read, and then re-read, Dale Carnegie's *How to Win Friends and Influence People*. This book will teach you the skills you need to work with people to accomplish team goals. Second, read *Switch: How to Change Things When Change Is Hard* by Chip and Dan Heath. And third, find a way to catch yourself when you feel yourself about to engage in a fight. Some carry an object around in their pocket that reminds them of who they are in their nobler moments. Some ask their team to call them out when they see their anger rising. Some simply take a break so they can step away from the situation and regroup.

Whatever you do, do not lose your temper, say inappropriate things, or fly off the handle. If you do that, you have lost. You will suffer the consequences of your own making. Whatever your approach becomes, remember this observation from Sun Tzu: The supreme art of war is to subdue the enemy without fighting.

The best generals win the war without fighting, just as the best leaders get the most out of their team without conflict. Learn from Patton's example: No matter how important you think you are; you cannot slap a soldier.

We would all do well to remember that anger is most often rooted in insecurity or fear. We need to address those fears by communicating openly with our team. Usually, a superintendent becomes angry because he or she fears the loss of control or lack of success on the project. This comes from trying to carry the burden of the project alone. If, instead, that superintendent would be transparent about the issues that arise on the project that are creating the feeling that the project is headed out of control, the load could be shared and the fears could be alleviated. Other members of the team could take on tasks to get the project back on track, or at least be made aware of pending problems so that they don't contribute to them. But the builder must ask for help if he or she is to leverage the power and strength of the team.

Everyone has problems. Every project has problems. No person achieves 100% of their potential if they try to stand alone.

We are all different and have different strengths and limitations. We must get past our insecurities about asking for help, because a top player is not one who functions alone, but is instead, one who functions well as a self-starter with the support of their team. The great builder is one who leverages the strengths of others and, in turn, contributes his or her strengths toward the success of others. That is how we all function at our maximum potential. That kind of leader is one whom everyone will want to work with, and whom everyone will follow.

Principle 8 - Be a Builder Who Learns from Good Experiences

The best builders are always learning. Those who are questionable are the ones who act like they are a victim of circumstance on their projects and always have excuses for why they do not go well.

We need people who will not allow themselves to be the victim of circumstance. Not only is this important in terms of performance, but it is also crucial that our people have good experiences and successful projects so they do not get comfortable with mediocrity. That would be the greatest failure.

It is one thing to have a project go wrong from time to time. Sometimes, that truly is beyond our control. It is quite another thing to have all the project team members leave a bad project thinking that situation is normal or acceptable.

A common saying is that "we are successful when we learn from our failures." But we also learn just as much, if not more, from our successes. When we have a good experience, we should take the time to reflect on it and identify the steps necessary to repeat it. It has been said that "practice does not make perfect, perfect practice makes perfect." That is true.

Good builders are not good builders just because they build. They are good builders because they have experienced success and can replicate it. They deliver results. When something starts to go wrong, they either take charge to fix it or raise their hands high and long until they can get help to fix it.

We discussed mental setpoints when we talked about jobsite cleanliness and the flow of materials. The message was that people will be as successful at maintaining cleanliness and inventory levels as they have decided to be. Whether they face success or setback, their jobsite cleanliness and control of inventory will once again meet the expectations they have set for themselves.

The inner standard for good experiences is also set by a mental setpoint. Successful superintendents have their mental setpoint positioned at predictable, organized, and successful. You may hear these people say, "What do you mean if this project is successful? It is going to be successful. We are here aren't we? It is our decision. Of course we will be successful." People with that outlook just do not know any other way.

We need to remember that we control almost every aspect of a project. We have access to the skills and materials we need to fix problems as they arise. When we find the quality of our work experience moving away from our setpoint, we shouldn't waste time complaining. We should act to fix things and bring the quality of the experience back to the level we expect. If you do not have a high setpoint for excellence, start actively paying attention to the amount of waste, failure, and mediocrity on your projects. Notice it and put a stop to it when you see it. Raise your setpoint. Let yourself be annoyed by it and go fix it. Implement change immediately; get creative if you must. Say to yourself, "If no one else will fix this, dammit, I will." Then go fix it, and brag about your job well done so others realize they need to raise their own setpoints.

When you find yourself consistently having good experiences at work, don't get complacent. A builder who is constantly learning is priceless. Those who do not tour other projects, learn from other people, or share lessons learned are missing an opportunity to be better, safer, more successful, and more helpful as a team player.

Continual learning makes the professional. Strong builders read books and trade publications to learn new information and improve their standards for best practices. They have insight into their weaknesses and reach out to mentors for help. Strong builders see problems and try to fix them with all the resources available to them. These builders visit other projects, learn from other teams, and open their projects for others to learn from them.

Principle 9 - Deliver Results Without Excuse

We cannot make excuses. We need to see things as they really are and deal with them efficiently to deliver the expected results. Results take care of our clients and the company. Results get our people home safely to their families. No one cares about excuses, and we must not excuse the signs of a project that is in trouble.

The previous principles are all intended to provide you a new framework of thinking to avoid the pitfalls into which unsuccessful people have fallen. This foundation for success is well-known to us, and it is time we built upon that foundation and discard old and unsuccessful practices and beliefs. The principle of delivering results is to not make excuses. We need to know what is not okay and move forward in a way that corrects it. Here are a few things we should not excuse or tolerate:

- **Bad worker morale.** This is most likely the result of our workers not feeling respected. We need to show our respect by providing them with the tools and materials they need, proper bathroom and break facilities, and clear, respectful communication. Our workforce will produce work at a level that matches their morale.
- **Lack of safety.** There are no gold ribbons for trying with safety. Lives literally depend on superintendents making sure that safety protocols are always followed.

- **Uncleanliness.** Our projects should be nearly perfectly clean. Morale, safety, and project costs are all affected by the cleanliness of our project sites. If we are off track, we need to clean up. Not being clean is a sign of lax leadership. Raise your setpoint and hold everyone accountable.
- **Lack of organization.** No one can manage a project effectively when it is mired in chaos. The days of the superintendent running around putting out metaphorical fires caused by disorganization are over. Create and clearly deliver the plan for your worksite and schedule and stick with it.
- **Insufficient communication.** We can no longer keep people in the dark. We need to use effective methods to communicate information to everyone on-site. When you prepare for a meeting, huddle, practice, form, or inspection, ask yourself if the information is formatted and communicated in a way that the recipients can understand. Everyone on-site should see as a group, know as a group, and act as a group. This can only come through good communication.
- **Unbalanced lives and poor health.** We need to stop paying for project needs with the health and mental wellness of our workers and their families. We need to send people home when necessary. We need people to go to their doctor's appointments. We need workers coaching their kids' basketball teams. We need to stop tolerating this unbalance. No success on the project can compensate for the damage done to marriages, children, health, or mental wellness. There is no excuse to run a team into the ground or take them out of balance.

Principle 10 -
Strive for Perfection

The principles described in the previous sections can seem overwhelming and impossible, unnecessary, or over the top. They are not. They are all possible. The problem is not that we are unable to clean, organize, and stabilize our projects; the problem comes because we do not have clean, organized, and stabilized minds and hearts. We need people striving for perfection. If we expect and strive for anything but perfection, we will get mediocrity. Only when people know that we expect perfection will they get out of the normal, mediocre, and disgusting state of our industry.

Think I'm being a little judgmental? I don't see it that way. We are all on the same team. We are at war with waste. It is up to us all to act against it. You have the right to expect more, and you should. You are in a professional position and have every right to expect high standards. Construction work as an industry can be disgusting, and we need to change that. If you do not want to change it for yourself, change it for others. We are here to challenge you to be excellent. We are asking you to take a stand and make a difference for the safety and well-being of the workers we are all responsible for. You can do it, but it will only happen as you strive for perfection.

Toyota is more profitable and successful than any other car company in the world because they expect excellence. Japanese culture as a whole already emphasizes the importance of striving for excellence, and Toyota's corporate culture was able to channel that tendency into an efficient workforce that aims for excellence. We need to buck the trend to be satisfied with mediocrity and create a culture of striving for perfection.

Principle 11 - Muda, Muri, and Mura

Muda, muri, and mura are words that represent the concepts of waste, overburden, and unevenness.

Muda means wastefulness. It can be actual waste on a project site, but it can also be non-value-added work in a process. These non-value-add activities are not needed for the customer experience or to deliver the project to the customer successfully. Waste can be re-work, unnecessary movement of materials, or any other activity that is not needed for us to deliver value to the customer. There are eight categories of waste: Transportation, motion, excess inventory, waiting, overproduction, over-processing, defects, and not using the wisdom of the team.

Muri means overburden, or excess overburden. This happens when workers are made to work too fast, or a floor is stocked with too much materials, or when a project management team has to process too much paperwork. Overburden occurs when a resource is utilized over 100% of its reasonable capacity.

Mura means unevenness, non-uniformity, or irregularity. This, in construction, can be described as variation in the flow of work or the variation in the resources available to do the work. Mura causes waste because it causes variation. For example, if there is an unevenness in the delivery of materials to a project site, that may cause excess inventory or waiting--which are ultimately waste.

It is a best practice for a superintendent to have a perch or lookout from which he or she can observe the movement of the project and seek out the 8 wastes in construction anywhere they can be found. Ohno circles--circles drawn on the ground where the leader can stand in vigilance to watch operations--may be utilized. Project walks can also help. If you can put observation and scrutiny into careful practice, you will be able to identify ways to systematically remove muda, muri, and mura on a daily basis. Projects do not only fail because of project management, finances, administrative burdens, and outside external conditions. Projects fail when they are subject to waste, overburden, and unevenness. A master builder should always anticipate these three imperfections on the project and take immediate action to eliminate waste, overburden, and unevenness through standard systems.

It is the duty of the project superintendent to create stable environments where all workers can succeed. This is the key to productivity and efficiency among all systems on the project site.

Part 3
Develop Effective Habits of Leadership

Now that we have reviewed the elements that build a foundation for success, it is time to discuss some of the key habits that will help a superintendent succeed on a project. The concepts may seem basic, but they will prove helpful to you in your role as you apply them.

Principle 1 - Hold Effective Meetings

Meetings are held for the purpose of communicating--taking information from concept to design to the workers. When you hear people complain about meetings, it is likely because they value their personal efficiency more than overall efficiency or perhaps the meeting was simply not run properly. Both issues can be effectively addressed.

First, let's consider people who may not like meetings because they value individual efficiency. When someone says they are too busy to attend a meeting, they are focused on their work only. When they say meetings are not necessary, they are likely planning to keep the knowledge they have in their head and work things out in the field or by themselves. Why does that happen? Two possible suggestions would be their desire for control and credit. They want control of the distribution of information or they want credit for carrying out the work themselves. This is a taker mentality. What we need on-site are givers--people willing to give of their time and knowledge to help the throughput of communication.

We constantly hear of superintendents who say, "I don't need to communicate the schedule or hold site huddles with trades because I have the plan in my head. I will just communicate with everyone on-site individually." We do not need superintendents in the field with the plan only in their heads because we are focused on the throughput of communication, not individual control or individual credit. When superintendents keep things in their heads instead of on a visual plan, they cling to control and crave credit and waste valuable time. Trade partners on-site are confused, without direction, and not productive because they are waiting for the superintendent to come save the day with information that only he or she has. It is destructive and unhealthy to our working relationships with our trade partners. It has worked for so many years because it's simply been the status quo. The alternative is a well-communicated plan and schedules shared and coordinated in meetings.

The second reason people avoid meetings is that they do often waste time because they aren't run effectively. Maybe the head of the meeting lets people wander off into personal stories or off-topic tales designed to boost the ego of the teller or the boss. Maybe someone feels it's the perfect time to air grievances that should be taken directly to the superintendent. Often, meetings are run with agendas so boring they could double as a prescription for a narcoleptic. None of these scenarios are acceptable.

Meetings should be effective, engaging, and upbeat. They should almost always have healthy conflict in the form of honest discussion about important issues relating to the project and they should keep everyone's attention. Well-run meetings are amazing!

So how do we hold an effective meeting?

First, have a clear purpose for the meeting. What do we want people to know or feel after it is done? If the meeting has no purpose, cancel it. If someone else asked you to hold the meeting for them and there is no apparent purpose, cancel it. If you hold a meeting and the attendees aren't prepared to present their information, then reschedule so everyone is prepared. You must be aiming for an outcome. "I want people to know the exact steps to getting the air on for this building," or "I want all of us to brainstorm on the root cause of this accident, so we can leave here knowing how this will never happen again" are both examples of outcomes that would require a meeting.

Second, prepare beforehand. Only the most skilled leaders can lead an impromptu meeting, and even then, it is an irresponsible habit. Be sure that the purpose of the meeting is clear to all who will attend and provide all attendees with an agenda that lists discussion points and who will present relevant information. Attendees will then know what topics will be addressed and won't be tempted to interrupt discussions because they are worried that their concerns won't be covered. Also, they will know the order of who will speak and who can leave when meeting topics no longer apply to them.

Having an agenda gives the leader of the meeting time to prepare to facilitate discussion and be aware of subjects that may invite off-topic comments. Leaders must always be prepared to redirect discussion that has strayed from the agenda either by inviting the wayward speaker to speak with them individually at a later time or asking them to hold their concerns for the next meeting.

Be prepared. Remember, the people in the room are there at a considerable sacrifice and cost to themselves and the company. Make every second worth their time.

Third, hook them from the outset. You must catch everyone's interest and get their thoughts flowing within the first few minutes. Consider creating a moment that everyone in the room will remember--a joke, a video, a call to action, or something else that will stick in their minds. Keep it relevant to what will be discussed. Get them energized and ready to tackle the topic at hand. How you start is likely how you will continue. Get their attention and focus immediately.

Fourth, keep their attention. Do not guide the meeting based only on content. Base the flow of the meeting on the energy and attention of the room. If you are losing people, speak louder, faster, or in a more interesting way. If they are bored because the topic has been thoroughly discussed, move on. Ask a question and get them talking. Whatever you do, don't lose the group and don't lose control. If someone is outright distracting or disruptive, call them out and steer the conversation back to the topic. Do it politely, but do it. Wasting the time of the group is disrespectful. An artful master of meetings will keep people's attention and think on the fly so that the discussion does not lose momentum.

Fifth, be sensitive about people's time. Let people know in advance how long you think the meeting will last. Then, if the meeting is done, end it. Do not wait for the initially planned for end-time to arrive.

Remember, we are trying to create a remarkable experience in our meetings, and nothing dulls that more than being stuck in a conference room when you could be acting on the information you've received.

And finally, provoke conflict. If your team is not having healthy discussions at meetings and disagreeing with each other, then they do not trust each other. We need to mine for open discussion in every meeting. If everyone is just agreeing and not engaging, then they will only remember a small fraction of the information conveyed, and we will have bottlenecked the throughput of communication. Get them talking, arguing, offering solutions or problems—whatever it takes to get them to own and process the plans and information presented in the meeting. Remember, everyone needs to weigh in and buy in. If you can host effective meetings, you can truly lead.

One of the first signs of a project in trouble is a job that does not have a regularly scheduled team meeting with people who show up on time, ready to participate. This is an immediate indicator of a lack of communication and an increase in waste and variation. Meetings not held regularly indicate mistrust, unhealthy individualization, and chaos. We meet to communicate information from start to finish during the project. We cannot communicate one-on-one with 300 people on-site without sacrificing time, money, energy, and morale. We must have effective meetings and communicate in groups. Effective meetings can organize chaos into stability, build relationships and trust, and ensure everyone weighs in and buys in. This is how we build the culture of our worksite.

On-site meetings should be hosted for the planning of the project, creating the plan for the following week with contractors, developing a plan for the next day, huddling up in the morning with all the workers, and filling out pre-task plans to prepare for the work. This all happens as part of the flow of the workday. We want to see as a group, know as a group, and act as a group.

If you coached a football team, would you make plays, tell a few people, and then do nothing else until game day when you walk on the field and begin to individually coach people mid-game? That is exactly what we do on projects sometimes. We make our plan in a vacuum, tell a few people, and do nothing until the day when we go around and talk to people individually so we can feel important for being needed. When someone says they are too busy to have a meeting, they are taking the easy way out and inconveniencing hundreds of other people.

Principle 2 - Have a Presence in the Field

Professional baseball catcher, Yogi Berra, once said that "you can observe a lot by just watching." One of the first priorities of a superintendent on-site is to be a presence in the field. Safety and quality cannot be managed by sitting in an office. Conversely, we cannot manage only from the field. There must be a balance.

Typically, we see assistant superintendents and field engineers taking the primary load here, but it should be balanced throughout the team. If someone is always watching the field, we will be successful on projects and ready for emergencies or situations that need correction. We appreciate the time spent by assistant superintendents in the field, which allows the lead superintendents to have time for planning and preparation. We also appreciate the entire team doing their part, especially for late days and weekend work. Project teams that spread out their presence in the field and divide the overtime hours among each member can be extremely successful. In no case should this responsibility fall to one person.

Achieving safety and quality requires supervision. These elements thrive when people can see and identify problems early. In Japanese manufacturing plants, they have what is called an andon which is a button, lever, or signaling device that will shut down the production line if problems occur. In order for the system to be effective, someone with the authority to stop the work must always be observing the worksite.

We may not have an actual andon installed in the field to serve as an alarm, but we need to apply the same concept on our projects. We need people to watch the line of production on-site and signal if anything is unsafe or defective. Our industry incentivizes production, not quality, and sometimes, not safety, so this kind of accountability—not policing—is necessary. This is where the assistant superintendent and field engineer can really make a difference. Many owners have commented in the past about how much difference field personnel can make if they are watching and controlling the quality and safety on-site.

A key component in this approach is the mentoring and coaching that happens. Workers should not be policed; they need to be coached and held accountable. Every worker on-site should be taught that they can signal the andon. There is no substitute for the one-on-one relationships that can be formed between the management team and the professional craft in the field. It can be a major driver in all aspects of the project, but most importantly in morale and the trust that workers have with leaders. If these critical relationships are not formed and if communication breaks down, that is when projects begin to have problems, morale declines, and workers begin to disrespect safety, quality, cleanliness, and every other key indicator on-site. We must build relationships with the workers. These relationships are the healthiest when founded in accountability and standards of excellence, not indulgence and leniency.

There is a concept in lean management that says we should go to the "gemba," or place of work. Lean leaders do not manage projects by sitting in the office and looking at charts and numbers all day. Effective and excellent managers run projects with numbers, reports, and visits to the worksite. On location visits allow us to see for ourselves what needs to be done.

Earlier in this book, Ohno circles were mentioned. If you were to work at a Japanese manufacturing facility, you might find yourself in a circle painted on the floor in the middle of the plant with instructions to stand in that circle and keep vigil. Your assignment would be to see—really see—what is happening on the worksite and make improvements.

Many find their first day of this is wasted. After some coaching and a little scolding, you would find yourself in the circle again the second day. This time would be different. You would see movement. You would see people and how they do work. You would see interruptions and waste. You would see the root causes of frustrations. You would see what you had never seen before when you were too busy to pay attention and actually observe the work. Then, and only then, when you see what needs to be seen, is it possible to mentor, train, fix, find root causes, and improve the overall workflow. Then you would know when to hit the andon for safety or quality.

Are you a manager on-site? Do you take the time to climb the tower crane and watch the flow of the project? Do you wait and watch long enough to see how things are going? Do you really observe the quality of work and safety? Do you look for safe behaviors, attitudes, habits, crew composition, and the quality of foremen on-site? Do you make sure that someone is always out in the field watching and covering the site? If so, great. You are already doing what excellent builders do. If you don't, then begin immediately. If no one is watching, then no one is there to signal the andon and stop the work when there is a safety issue and prevent someone from getting hurt.

Principle 3 - To-Do Lists and Personal Organization

Being a great leader and builder comes from being organized and focused. People who do not take the time to develop personal organization systems can find themselves in the field fighting fires because that is the only way they know how to operate. Conversely, being able to plan and focus on a role will allow us to organize large efforts ahead of time and execute them effectively. The way to do this is to know our role, schedule leader standard work, and control our to-do lists in a way that keeps us in control. The following suggestions will help you in creating your personal organization system.

First, you must never forget what needs to be done. There is an old saying that a dull pencil is better than a sharp memory. This is absolutely true. With hundreds of thousands of material components coming together at specified times in a project, it is not wise to rely on our memories. We must have a to-do list and a system to utilize it. Remembering is the difference between being a poor manager and an excellent one. People who forget things are a liability. They are the ones who cause other members of the team to stay up at night worrying.

The trick is to write down everything in one location—commitments, deliveries, meetings, ideas—and reference it regularly. Do this to stay on task and be effective. Once you have a repository of items that must be done, it is key to sort the list daily and focus on priorities.

When organized leaders arrive at work they sort through their Outlook, reference their to-do list, and prioritize items that need to be done that day. That leader can then add that list to the calendar day and go to work according to the plan. People who have developed this habit are under control, steady, happy, on-time, and often promoted. They control their time; it does not control them. They protect their plan and drive success.

Create your leader standard work plan. Leader standard work is when you transfer your tasks to a written plan so you can carve out time for the standard activities that must be done to carry out your role. For instance, a lead superintendent might have items on his weekly plan such as walk the jobsite, study the drawings, update the schedule, take a reflection walk, and attend the OAC meeting and foremen huddle. (Notice, there is no mention of items like "boss people around" and "micromanage.") Once this leader standard work becomes fixed, include it in your Outlook calendar. The tasks on the leader standard work plan should be protected from non-value-add meetings and other distractions.

To-do list items can fill in the rest of the time in your workday, but you should deal with your to-do list in order of priority. If something is urgent and important, do it as soon as possible. If something is not urgent but important, schedule a time to do the work. If it is urgent but not important, delegate it to someone on your team who can handle the task. And, if it is not important and not urgent, delete it and do not think about it again. In that way, you will accomplish the most important things first.

Imagine someone filling a jar with sand, then gravel, and then rocks until it overflows. Then, imagine the person starts over, this time placing the rocks in first, the gravel second, the sand third. What overflowed the jar before, now all fits.

Our time works the same way. Fit the most important things into your schedule first and do the rest in order of priority and you will find that you have time to do everything that is required of you. Schedule your personal and family time first, then fit in your leader standard work tasks followed by your properly prioritized to-do list items, and you will be adding rocks, then gravel, then sand to your daily time jar. You will have prioritized best, better, and good in their proper order and the most important items of the day will all fit.

Remember, discipline will always beats mere talent. Discipline yourself to never forget what needs to be done and the order in which it needs doing

Principle 4 - Provide the Four Motivations People Need to Be Effective

People need four motivations to be effective at work--a place to work, tools and equipment, training and instruction, and time to perform the task. These may seem commonsensical, but you may be surprised by how few people plan for these things and how much time is lost as a result. By not allowing the throughput of communication, understanding, thought, and action that each person needs, we end up with waste and variance. To be effective, we must ensure that these four basic needs are met for all the workers on-site.

First, provide an appropriate place to work. Do not underestimate that everyone needs a place to organize themselves, think, and do their work. Each of your team members likely needs a desk, enough space to work with drawings, screens for the computer that allows them to expand content for understanding, and a place where they can sit and be comfortable and productive. All of this should be in an environment that facilitates communication, collaboration, trust, and (at certain times) fun.

For team members, providing an appropriate place to work may be a hard concept to plan for and understand because it is not easily measurable. The effects of bad planning in this area are not easily seen, but common sense and observation tell us that providing an appropriate place to work is very important. A team should thoughtfully plan the height, size, and configuration of desks and the size and quantity of monitors based on the available layout within the trailer. The trailer's design should benefit and support the way people need to work, act, and feel.

For crews, workers need a place to put their lunch, their tools, and their plans. The purpose of gang boxes is two-fold: a space for tools as well as support for the productivity of the worker as they prepare their day and do their work effectively. The criterion should not be, "How much does this gang box cost?" or "Can they get by with this?" but rather, "Does this gang box, station, or area support the work of the craft and enable the crew member to plan, build, and finish their scope?" On your projects, do not underestimate the need for people to have a place to work so they can be effective. Additionally, tools located in arbitrary locations result in treasure hunts for the worker. It is very effective to have tools in secure locations that are organized, repetitive, clean, and sometimes shadow boarded.

Second, every worker needs tools and equipment to do their work. We need to purchase, organize, and maintain tools and equipment so workers have what they need. Historically, we lose money with high-risk exposures and wasted man-hours. Only through waste, loss, or overspending can we lose money on small tools and equipment that help workers be productive and safe.

It is discouraging to hear of supervisors who question and halt the purchase of needed tools for the field or office. The halting of a simple two-hundred-dollar purchase may hold up countless man-hours or productivity and comes at the expense of quality, cost, or someone's home life. Even if only five hours of production is lost in the course of a project, that still equates to well over three hundred dollars in loss to the company. If your workers need tools or equipment to effectively do their job, they should have them.

The third motivation needed to be effective at work is training and instruction. If training our people is expensive, consider the cost when they are not trained. Our success and effectiveness is directly correlated with training. Everyone needs training in safety, quality, lean, and the specific work they are doing.

Training should be intentional. For instance, creating a list in the Team Weekly Tactical meeting that shows training events and times for team members can be very effective. Run your project in such a way that everyone can attend their training and the team can cover for them when necessary. The biggest insult to our people is to hold them up at the project and cause them to miss their training because leaders were not intentional enough to plan for coverage.

Train, train, and train some more. Make training more effective. Choose better topics. Have people report on what they learned then encourage them to train others. Double down on this and challenge your team to improve themselves daily and weekly. How are millionaires, Nobel-prize winners, and legends of history made? Hard work, constant learning, and a determination to self-improve every week. Better yet, let's expect those key elements from all our people daily.

Fourth, workers need time to perform a given task. We are not productive when workers too quickly perform a task due to time constraints. Fast work on-site may mean fast planning, additional resources, staggered hours, prefab, and a number of other things, but it does not mean individuals working too fast. It also does not mean entitled workers going slow because they are not accountable. What is needed is a steady and effective pace.

Give your workers realistic task durations and accountability. If working at a fast pace creates the potential for a worker to become injured, we need to point out that potential for injury, slow the worker down to help them create an effective flow, and increase the support around them. This concept should be immediately implemented on our projects. It is not requisite to ask workers to pay for the sins of someone else who didn't do the job to the best of their ability. If there is a design change, that change should pay for the resources to fix it. If there is an item we want to absorb for the owner, we should do it out of our fee, not the health, well-being, and safety of workers and the family time of management. We need to allot everyone the time to do their work.

Leader or Worker? - A Story from the Field

One time I was in Austin, Texas, thirty feet down in a limestone hole helping build a concrete basement for a Whole Foods World Headquarters. I was working with the reinforcing company in several areas of the project and trying to make pour dates for decks, columns, and walls. I worked with one foreman who was always late, behind, and disorganized. Another foreman with the same company was always on time, on top of quality, and very effective. Being a young kid and wanting more experience, I asked the good foremen what the difference was.

He said, "Jason, do you see that other foreman? He is always working with his people. He always has his head down working."

I said, "So? What difference does that make?"

"Well, Jason, his men never know what to do because he is working and not leading. You see, I spend all day getting material and instructions to my people, and they do the work. That is my job and the only way I can keep everyone working well."

That lesson has always stuck with me. It was the first time I realized it is our duty to get information and materials to our people. Our job is to prepare so that others can do their jobs. It is not to look like hard workers. It is to get others the tools, supplies, and information they need to do their jobs effectively.

Principle 5 - Be Schedule Driven

A superintendent is schedule driven. This seems like common sense, but what might come as a surprise to you is just how much the superintendent must know about scheduling and how often he or she will use that information. A superintendent should always be part of the creation, maintenance, and implementation of the schedule. This will guide the success of every other thing that superintendent will do. A super is the timekeeper on the project. He or she keeps everyone accountable—in a kind, professional way—to the time component of the project. There are also responsibilities for handling the components of "how, where, and what," but the "when" is the main focus for a superintendent.

Project managers with successful projects should be able to affirm that the super "kept us accountable and moving forward throughout the project in a healthy way." There are no successful superintendents with "woe is me" or victim mentalities when things go awry. Superintendents plan, organize, drive, rally, and win. They help set the rhythm, help others match the rhythm, and keep the rhythm going on-site.

When creating a plan for the construction project, a superintendent must first know the constraints of the project and visualize a possible flow. Everything on a project—all one million pieces of it—must flow together in a seamless stream to assemble at the right time and in the right sequence. Identifying this allows us to identify our plan.

The plan is the predecessor for everything else. Safety planning starts with the plan. Costs are affected by the plan. Quality is implemented from the plan. Production is achieved according to the plan. It all starts there. Safety, quality, schedule, cost, and production are all equally important, but it all begins with the plan. It all begins with the schedule. Therefore, it all begins with the superintendent.

Once there is an established plan, it is the lead superintendent's job to maintain flow, keep project energy high, feed the project with materials and manpower, and shield the work from variation. The plan will change but the skilled superintendent will have plans A, B, C, D, E, and F. He or she will always know how to be nimble in a flow. A superintendent will always know how to respond, react, and maintain morale on-site and know how to meet the next milestone. There really is no need for a superintendent who doesn't have a schedule.

Principle 6 -

Study the Drawings, Talk to the Building, Predict the Future, and Share What You Learn

There are four key habits that are common to most successful superintendents. We study the drawings, do reflection walks, use our schedule effectively, and communicate what we learn. These are universally helpful for any position or personality type.

First, take the time to study the drawings. This is at the root of what we are paid to do. Schedules are made from drawings; budgets are then made from the drawings and schedule. Quality and safety are planned around them as well. Everything starts with our ability to read and understand drawings. Even design management is the act of trying to collaborate and get to a set of drawings. Construction drawings are essential to what we do, how we do it, and why we, as a general contractor, are paid so much. We can understand the design intent, build the project visually in our minds beforehand, and create something from nothing—no one else can do that. That is our skill set. As such, it comes as a surprise that many would learn this valuable technique only to later ignore it and suffer from low quality, poor planning, and unsatisfied customers.

We need to study the drawings daily to know what, when, how, why, and where we are building. Try to set aside at least a half hour each day to do this. If you cannot get one half-hour, try to get fifteen minutes. If you forget one day, remember the next. Schedule it on your calendar. Focus on the areas you need to study just ahead of where the project is now. Study details, dive into the hard areas, and make it fun. Highlight, mark, and take notes—get addicted to it. We should drool when we see a nicely printed set of new drawings. We should "go down to Staples like it's Black Friday" (to quote Buddy Haws) and empty out the store in an effort to get the highlighters, pens, stickies, markers, and other supplies we need to help us fully understand and interpret our project drawings.

A great superintendent will study the drawings daily. It is what we get paid to do. Without this habit, we can become brokers and construction managers instead of builders.

Second, build time into your day for reflection walks. Reflection walks are when you go to the place of work and try to see every aspect of the site. Try to talk to the building and ask it questions. Try listening to it. It will tell you where to go next and what to do. The building will tell you where the problems are and call your attention to the things you would not otherwise see.

Now, some of you may think I am heading down a metaphysical path with this whole "talk to the building" thing, but I am not. The building will talk to you. Just listen. Don't discount it until you try it more than once. Superintendents who take the time to walk the building will always know where to steer the ship with the team. They will always know what to do, especially after studying the drawings and seeing what the building will grow to become in one week, one month, or one year. This is how we go from okay to excellent. Very few leaders are found behind schedule or with a project in trouble who take the time to do this.

You will also be able to see behaviors while you are on your walk. You can snap pictures, set reminders, and send messages about upcoming efforts. This can be reported the next morning in the worker huddles on-site. There are really no words to describe the magic of reflection if done in a quiet, focused, and clear-minded way. Designers may know what the building wants to be, but we talk to it daily as builders.

Third, be in the schedule daily and predict the future. Great superintendents can see the future. That is their job. A project manager should read minds and the superintendent should be able to see the future. There is nothing quite like a superintendent who creates, owns, updates, and manages his or her own schedule. It literally becomes the crystal ball through which we see the project and the future. Without it, a superintendent can become disconnected with the future and stay only in the present. That will only lead to fighting fires. Every superintendent must avoid—as much as possible—someone else completely creating, owning, and updating the schedule. If it must be delegated, proceed cautiously. If it can be avoided, avoid it.

Taking the schedule away from a superintendent would be analogous to taking a hammer away from a carpenter and asking him or her to still do their job. If a superintendent will study the drawings, walk the project, and then be in the schedule, they can send reminders, prepare work, get others to see the future with them, and leverage their efforts. It is powerful when you get daily little email snippets of the schedule from the superintendent after he or she has studied the drawings or done a reflection walk.

Fourth, communicate what you learn. After a superintendent studies the drawings, walks the project, and schedules effectively, he or she will realize things that need to be scaled to the team. It may be snippets from the schedule, a message, or another brief way of communicating. But it may be more than that. There may need to be a highlighted drawing made for tomorrow. We may need to update the exterior sequence. We might want to add a few notes to meeting agendas to communicate. We might want to make a five-page drawing to communicate the plan for the next five days. The point is this: the plan does little good in the setting of a team if the plan is only in the superintendent's head. A plan is only as good as it can be effectively communicated.

Principle 7 - Problems are Not a Problem

One of the main concepts that a superintendent should understand is that problems are not a problem. Every project and every person has problems. The strategic advantage is to raise all problems to the surface by creating a stable environment with flow and then to systematically and passionately remove those problems on a daily and hourly basis. One of the best ways to implement this system is to memorize the 8 wastes in construction and have the entire project memorize the 8 wastes. It is helpful to create the kind of culture that showcases how and why the 8 wastes upset the common group so you can point out how you can remove those wastes on an hourly and daily basis. Any roadblock, whether it be big or small, should be raised to the surface for removal. This will create flow, ensure quality work is installed, maintain safety, and allow us to meet our schedules. Thinking that you do not have problems, will leave the problems in place, not allow us to fix them in enough time, and it will interrupt the flow of work causing waste and variation.

Remember the Ohno circles? The purpose for standing inside the circle is to observe and learn about the movement of the jobsite to see where improvements may be made. The designer of the Ohno circle, Taichii Ohno, is known as the father of the Toyota Production System and also inspired Lean Manufacturing in the United States. If Ohno asked his workers if there were any problems seen from the perch of the Ohno circle, "no" was the last response he accepted. The following quotes are attributed to Ohno and it is best practice to adopt his philosophy: "Having no problems is the biggest problem of all" and "Progress cannot be generated when we are satisfied with existing conditions."

Principle 8 - Implementing Lean on Projects

There's a specific order in which a superintendent or project team can implement lean on a project. The specific order can be listed the following way:

Setting up contracts for lean

All work orders, exhibits, or contracts should list intentions for lean operations and include items such as wall coordination, material storage, concepts like "nothing hits the floor," morning huddles, 5S-ing, and any other lean principle which should make its way into a trade partner's contract from the very beginning. If this is not done, history shows that subcontractors may find fault with the merits of such systems and attempt to back charge the general contractor at a later date. We need to be fair with all trade partners and be very clear about expectations throughout the project.

Winning over the workforce

Once the proper lean practices are in contracts, the absolute first thing a project team must do is to win over the workforce. The workforce must be treated with respect, and they also need to feel that respect. This comes from constant communication, modeling, and in maintaining respectful systems on-site. Bathrooms, lunchrooms, and the communication systems should all work toward an environment where workers feel respected. It is only in this environment that workers will ultimately feel a sense of reciprocity and desire to care for the needs of the project team and the project as a whole.

Cleanliness and organization

Cleanliness and organization cannot be overstated. Cleanliness and organization is an environment where everything thrives. Maintaining the attributes of cleanliness and organization shows great respect for workers, maintains a sense of appreciation amongst your entire project team, and will also enable you to sustain or broaden your existing culture and environment.

Sustaining the culture

You must find ways to sustain the culture you have created. The collective behaviors, beliefs, and actions of the team will create the culture you have on-site. As the leader, you cannot allow behaviors that are destructive to your culture. You have to shape the beliefs of your project in morning huddles and through constant communication. You have to make sure that everyone's actions are guided properly by the right visual systems, operational structure, and reinforced habits. It is your job to ensure that a proper culture is being maintained on-site. One of the best ways to sustain a culture is through standard systems but also to tour other people through your project and invite project workers to showcase the things they are doing to create lean cultures.

Balance and Stability

The balance and stability of the team must be maintained at most costs. If a project team, including trade partners and workers, feel overburdened or uneven, it will affect every operating system on-site. When a superintendent has worked to create a culture, and to sustain that culture, he or she must also protect the team from overburden and shield the entire team from variation and waste. This is the only way to ensure the maintenance of the environment you have created. Everything else stems from team balance. Your productivity, your safety, the quality of the work, the innovation and use of technology--everything relies on the balance and stability of the team.

Foresight and Planning

Foresight and planning must be done from the beginning to the end of a project. One day in pre-construction or planning will make up for a week in the field. One hour in pre-construction or planning will make up a day in the field. All work must be successfully planned and prepared. You, as the superintendent, must follow your standard daily habits to see the future, lead with foresight, plan work so that it can be executed properly, and be 100% prepared and ready for implementation.

Clarity and Alignment

Once there are standard and stable environments in the office and in the field, the direction of the project and the immediate and long-term goals must be clarified and frequently communicated to everyone on-site. Everyone should be aware of the contractual end date, the conditions of satisfaction for the project, and the immediate goals for the week or month. There should be no question about project standards and how the work is to be carried out. Changes, adaptations, major problems, or recoveries must be communicated to the team so that everyone can see as a group, know as a group, and act as a group.

Vision and strategy

With clarity and alignment of proper communication come strategies to overcome certain aspects or phases of the project. The vision for how to overcome mini-marches, accomplish milestones, or complete specific efforts has to be clear to everyone. Teams must be nimble and adaptable when difficult situations arise, but the strategy to conquer has to be known play-by-play by everyone on-site. Culture will eat strategy every time, but not if you have a proper culture, stable systems, plans, and communication on how to win. You can only strategize once you have stability. Form a plan and strategy with all trade partners on-site to accomplish specific tasks, goals, and to meet specific milestones. □

Principle 9 - Develop a Healthy Relationship with the Project Manager

The project manager and superintendent have equally important positions on any project. They can do more when they work in cooperation with one another than each would be able to do on their own. Though they are two people focused on different aspects of the project, they have a common goal of completing the project successfully. By working together they can leverage more oversight, diversity of opinion, and increased accountability.

The superintendent is responsible for keeping the project moving at a reasonable pace. He or she needs to anticipate problems that may arise and be prepared to quickly solve those issues. The project manager is responsible for maintaining a business mindset to reduce risk on the site and to interact with the owner to make sure his or her expectations are met. Clear expectations of each role will eliminate an adversarial relationship. Tension is best avoided if you maintain your role as supervisor and allow the project manager to fulfill his or her own role.

As a superintendent, support the project manager by showing an awareness of project finances as you stay within financial constraints by properly forecasting contingency plans. You can also support the project manager by making safety, quality, and timeliness a priority in the field. Make inroads with your project manager by supporting team health and making sure that everybody on the project feels like they are part of an effective and high-performing team throughout construction.

It is important to dutifully execute your daily, weekly, and monthly plans with the needs of the owner in mind. Maintaining focus on the owner will cause you to think in ways that will readily make the project manager's job easier and he or she will be more accommodating to you in return. If you are focused on maintaining safety and controlling costs, you show the project manager you respect him or her as a professional.

When the project manager feels respected in his or her role on the project, it is easier to reciprocate by finding ways to provide needed manpower and funding as you need them. It is never a good idea for a superintendent to demand rather than to give.

There must be a balance between the worksite and administrative offices. Both the project manager and the superintendent should cultivate awareness of one another's professional responsibilities from the beginning and maintain that respect throughout the project. Proper communication between the positions is essential for the success of both the superintendent and the project manager and, ultimately, the project itself. If those in the office are only worried about cost and risk and those in the field are only worried about schedules and manpower, then we are all at risk for mistakes, injuries, and other unintended consequences.

Our product is this: workers putting work in place in buildings. That's what we sell. Superintendents organize and deliver that product. Project managers support and enable the superintendents, act as advocates with the owners, and make sure that we are financing that product. We are not selling the superintendents' egos or the project managers' pride. We really need to understand that acting in tandem so workers can perform their jobs is the most important point. We don't want a hierarchy to enter into this dynamic situation. We do want to recognize the value of both positions. Working together, the superintendent and project manager can do far more than if they view each other as adversaries.

Principle 10 - Staff and Field Officers Should Support One Another

One common theme that arises from a study of military history is the conflict between staff and field officers. Typically, staff officers would work at headquarters located a safe distance from the battlefield. They would handle general logistics and coordinate the efforts with allies and other organizational tasks. Field officers would be the boots on the ground working with soldiers in combat and making decisions on a day-to-day basis. Often, these two groups of leaders would clash over what was the best course to victory and as a result, a feeling of dislike would develop between the two groups. Patton, Eisenhower, and Field Marshal Erwin Rommel all mention the division between staff and field officers in their biographies and stressed the need to merge those two worlds.

Likewise, it's common on construction projects to sometimes experience disagreements or to harbor an "us versus them" mentality among office staff and field staff. In reality, the office staff and field staff are part of the same team that owns all aspects of the plan.

Thinking in definitive terms of "office" and "field" is one of the most damaging and detrimental philosophies affecting our projects. If those in the field only look at schedule and will not focus on procurement, finances, customer service, and owner relations, then they are not doing their part to ensure the success of the project. If the office only focuses on procurement and finances, they will miss very important aspects of the schedule, flow in the field, and how to take care of craft workers. In other words, they will not be doing their part to ensure the success of the project. Everyone needs to own all aspects of the plan.

When silos exist between the field and office, there is also fear of stepping on somebody's toes. That sort of culture is divisive and not conducive to project success. Ideally, we want the superintendent to be able to ask anything he or she wants about finances, procurement, and owner relations and the project manager and project director should always be able to ask any question about schedule, field operations, quality, and safety. There should be no hesitation to ask nor should there be any feeling of defensiveness when asked. Everyone on the jobsite should be accountable to one another. There should be no situations occurring when people pull rank. We love when a field engineer can hold a project director, field operations director, or superintendent accountable with a comment or a challenge to do the right thing.

This does not mean that everyone needs to make everything their primary focus. The superintendent will focus mainly on the schedule. The project manager will focus mainly on finances and customer relations. But everyone should be aware of all areas of concern, weigh in on them as needed, and buy-in to all aspects of the project. There should be no divisions.

Construction is a team sport. We have clear lines and defined roles on a construction project; in fact, it's one key indicator of a successful project. But those defining roles, those lines, are thin lines and can be crossed when necessary. Just as a forward can step into a guard's position on a basketball court when needed, the project manager should be able to step in and help manage the schedule and run trades for the superintendent if required. If the project manager is absent, the lead superintendent should be able to step in and run an OAC, take care of owner requests, and help in other areas.

There should be a horizontally equal feel to the organizational structure of the team. If you were responsible for everything on the project, how would you act? How would you execute your day-to-day responsibilities? Focusing on that from time to time is important. Don't silo, disrespect others, or reduce the amount of communication that you provide. Recognize the value that all positions play on a project.

There are some key considerations that will help field and staff officers be successful together. One is working in close proximity in an open office space, not separated by walls, so that frequent communication is possible. The other consideration is communication itself that comes in the form of huddles, daily check-ins, and weekly team meetings that are meaningful and draw the team together. Also important are team building sessions where everyone on the project can develop rapport, get to know each other, and develop trust in order to foster good team health in all that they do.

Principal 11 - Short-Interval Scheduling Should Always Be Tied to Milestones

On some projects, the work is complex enough that the lead superintendent will work with a superintendent level I or level II. The lead superintendent is then in charge of planning and executing the communication structure and work of the entire project, while the level I and II superintendents will be focused on portions of the work.

These components of the overall project must be coordinated and executed according to their own schedules—a short-interval schedule. Level I and II superintendents are responsible for focusing on the details of these components for the overall project and will ensure that the work is done well, on time, and with the supplies and materials required.

Communication is essential between the lead superintendent, project manager, and the level I and II superintendents to make sure that the short-interval schedules are completed at specific points or milestones of the overall schedule. If they are completed in a timely manner, the overall project can continue to move forward, but a delay disrupts the entire construction schedule. Therefore, superintendents level I and II have a duty to ensure that they are focused on the details of their specific projects and know which areas are reliant on them to meet their short-interval scheduling goals to complete work on time.

Communication will prove invaluable when it becomes necessary to share manpower, materials or resources between the various aspects of the project assigned to assistant superintendents. By coordinating their efforts with the senior superintendent, he or she can ensure that needed manpower or supplies can be shifted as needed.

If other areas of the overall project are suffering from delays due to a shortage of materials or manpower, that is the concern of the lead superintendent. The focus of a superintendent level I or II should remain on the portion of the project they have been assigned to oversee. As long as they have the materials and manpower they need, they should move forward with their responsibilities.

The lead superintendent should be able to trust the level I and II superintendents to oversee the construction of the projects they have been assigned. They focus on areas of the project, not the project as a whole. The lead super should be able to count on the level I and II superintendents to complete their aspects of the project on time, without having to micromanage their efforts.

Assistant superintendents are to make sure that the day and week planning is executed well. Senior superintendents are to make sure the week, month, and year-long planning is going well. Senior superintendents are there to make sure that design, RFI answers and specifications, and any other needed information is there to support the short-interval plans of the assisting superintendents.

If something happens relating to manpower, materials, or delays on the aspect of the project you are supervising, it is essential that you reach out to the project manager and lead superintendent. Do not fear transparency. Trying to hide difficulties you cannot solve on your own will only lead to further delays for the entire project. Be transparent and open.

All levels of superintendent are important. Nothing works without drive in the field and energy of the team. We create that energy by having layers of oversight built into our leadership so that somebody is watching and providing accountability. We must have somebody motivating, communicating, and looking at the details or the specific tasks covered in short-interval schedules will not be executed. One level of leadership ties into the other. Differing roles are equally important. And they are all aligned vertically with project milestones.

Principle 12 - Always Make Production Tracking a Focus

Problems on a construction project belong to the team. We need to be able to see problems as soon as possible so that they can be addressed quickly and effectively to minimize delays to the project. No one should be trying to hide problems on the jobsite.

Problems can be found quickly through production tracking. A master schedule with short-interval schedules tied to milestones on the master schedule allows superintendents to quickly identify delays and shortages. It also allows them to predict future issues.

Everyone on the project should be anticipating roadblocks and potential problems as far into the future as possible so that we can continue to execute our plans daily with a consistent flow. The key to success is flow.

If we have a pull plan that ties to a milestone, we need to make commitments on a weekly basis according to that pull plan. Every trade partner on-site should make commitments as a team for the next day and the next week, then track those commitments with daily progress reports.

After tracking these commitments, performance should be graded. In lean construction, there's something called a Percent Plan Complete. PPC scores track the amount of activities that you've committed to and how many have been completed. For each contractor in any given week and for any phase of work, there will be a PPC score that will track the percentage of the things committed to do as well as actually completed.

Tracking progress in this way allows contractors to communicate reasons why they were unable to complete their commitments so we can help them make corrections. Also, this allows us to grade the overall performance and behaviors of our organization and our trade partners. We use a series of objective grading criteria that allows us to identify if a contractor is being successful. But the main benefit of the Percent Plan Complete is to track and remove roadblocks to completion. Fanatical roadblock removal is the key to any production tracking system on-site.

We need to gather this information regularly to create accountability. Quick, proactive reactions will allow contractors the empowerment to immediately widen their circle and tell the general contractor before the problem grows.

Every action on-site flows to a rhythm. If a roadblock interrupts the rhythm of a project because of lack of materials, manpower, or information, then it will disrupt the flow. It will make us lose money, put us behind schedule, or cause other unintended consequences.

As we track production, we make commitments to try to create flow. As we do so, we not only ask contractors what they want, we ask them what they need to complete their portion of the project by the scheduled milestone. Then we try to level out the workload or the flow of every contractor.

Sometimes contractors will declare, "I have to go faster (or slower)." We must then find compromises so that we optimize the schedule of the entire project. Together, we need to create a plan to meet the scheduled dates and find the win-win so that we can remove roadblocks.

We must be nimble. No matter how well we plan, issues will still arise. If everyone is committed to maintaining the flow of work, then we can react to these issues quickly, adjust our plan, and recover from any negative impacts sooner rather than later.

A superintendent must know the costs of an interruption in production that requires a course correction as well as other activities and teams the interruption will affect. Making production tracking a focus will help provide these answers so he or she can make good decisions to keep a project on track.

Part 4
Being a Good Team Leader

Principle 1 - How to be a Good Neighbor

Being a good neighbor is one of the main responsibilities of a project superintendent. We must protect our neighbors, the people around us, our customers, pedestrians, motorists, and anyone within close proximity of the project like we would protect our own family. It is appropriate to advise that we should treat the neighbor on the corner as if our own grandmother lived there. One of our main goals is to elevate the awareness of the team to really care about the needs of others. The project team only wins when we can stay on budget within schedule, with a quality project, with a team balance, in meeting everyone's individual career goals, and when we have delivered a remarkable experience for everyone that comes into contact with our project.

If you were to ask yourself who your customer is on the project, what would you say? Would you say the owner of the building is your customer? Would you say the end-users of the building maintenance team is your customer? Or would you say the designers are your customers? Any of these answers would be correct, but there are additions. Your neighbors are your customers, your trade partners are your customers, the people in the adjacent building are your customers, the vendors are your customers, and trade partners that go after another sequence of work is a customer. Everybody on your project should be treated in a like manner. This is the way a project can elevate performance and really take it to the next level to create flow and a good quality work product.

To really grasp the concept of customer service you must first focus on your extended network. For instance, if you work on a campus, you probably have customers from the facilities department and consequently, you interact with other departments within that campus organization such as marketing, life safety, inspection, engineering, and other functional departments. We cannot look at them as if they are there to fulfill needs for us. We must consider ourselves to be part of their organization and evaluate ourselves as one of their service providers. Do we treat them with the proper amount of courtesy and respect and consideration? Do their problems become our problems because we care to help? Are we concerned about their team and the streamlining of their operations in the flow of work? Do we care about their reputation? Do we know the things that keep them up at night? These are all opportunities for us to consider as we build a rapport of customer service.

We can create a remarkable experience on our project if we know who our neighbors are. We want our neighbors to see our management team and our project in a favorable light and as a positive participant in the community. There should be no negative impacts, no interruption to systems, no trouble, minimal noise; the roads should be situated and flagged inside and barricaded in a way that people can easily make it to their homes and their places of business. We should consider newsletters as a way to inform, lessen any frustrations, and encourage understanding regarding construction activities affecting their daily lives. Our neighbors should know that they are one of our top concerns. We should take them seriously and always have their needs be our first focus. This is often difficult because we have a tendency in relationships to feel taken advantage of and to feel as though we are always giving with no reciprocity. If we can get out of that negative mindset and have it as our mission to give and provide and take care of our neighbors, that is when things will become remarkable for us and when we will be true service providers for everyone we come in contact with. That is when we will truly win over the owner.

Putting the concept of great customer service into practice begins with the project team identifying their customers--their neighbors. Additionally, they need to come up with an actionable plan that pinpoints how the project team can be good neighbors and take care of all customers for the project. The project team should then develop steps for communicating regularly to all neighbors. Most importantly, the project team really needs to elevate their awareness of neighbors and really have a heart connection with taking care of them. The superintendent is the guardian of the project and is the guardian of the neighbors.

Principle 2 - Love the People and Inspire Them

You can always tell the difference between a superintendent who is just trying to do his or her job and a superintendent who leads with care and concern. The latter will always be the one who will have the most influence on the project. Now, we are not at work to make friends and close relationships that outweigh the need to do business and have respect and appropriate relationships on-site, but everything we do within those respectful and professional relationships should be guided by care and concern for the people on the project. Until we recognize that everything we do moves through and is about people, we will not be as successful as we can be.

We are not in the manufacturing business. Most people think that construction is about producing a product at the lowest possible total cost in the shortest amount of time with the greatest amount of value. I would argue that we are in construction to build people first who then in turn build great buildings and great things. We are a customer service business. We take care of the needs of our customer and all the customers around us. If we do not foresee the people who will eventually use our buildings, the people who will maintain our buildings, the people that are currently working on our buildings then we have a gap that we need to close in our own personal leadership. We need to act, control, and lead out of respect for those people and make them our priority.

The best superintendents in our industry are the ones who are approachable, lead with a vision of taking care of people, run safety because they want to keep people safe, and install quality work because they care about the end-users. These superintendents are not only approachable, but they are also authoritative and respected. They are the leaders who can get people to rise up and follow them because of their influence. At the end of the day the definition of leadership is influence. There are leaders throughout history who have led by influence but were not good leaders. Nazi leader, Adolf Hitler comes to mind. His influence and leadership was without morals and ethics. In contrast, President Ronald Reagan earned the nickname of the "Great Communicator" as he made positive connections with average citizens. A superintendent can have either good or bad intentions, but a leader with influence will be better suited if their intentions, track record, and experience prioritizes the safe keeping of people. The leaders in our industry who make that vital connection will always be more influential.

Find role models who exemplify this behavior then make resolutions for your own style of leadership. Focus on giving first and then intentionally practice being the leader that you want to be. Books like *The Go-Giver, A Surprising Way of Getting More Than You Expect* by Bob Burg and John David Mann and *The Leader Who Had No Title: A Modern Fable on Real Success in Business and in Life* by Robin Sharma give credence to the merits of true leadership. If you pursue an education in leadership style and successfully practice what you learn, you will expand your level of influence as a leader and also bless people's lives along the way. Givers gain and takers lose. Givers in this industry will make a difference for thousands of people. Takers not only suck the energy out of their own teams, but they give this industry a bad name, and they make working on their projects miserable for everybody.

Principle 3 - Exhaust Bad Behavior

There is nothing that will tear down the motivation and morale of good people on your project site more than watching you tolerate bad behavior. As the lead superintendent or an assistant superintendent on any part of the project, you cannot tolerate bad behavior. Remember this quote that has also been shared in other sections, "The success of any organization is determined by the worst behavior the leader is willing to tolerate." As a superintendent you must have standards, people on the project site must know what you expect, and you have to exhaust bad behavior.

Within your leadership on the project site, you should ask yourself this one question: *What are the behaviors on-site that I am tolerating but want to stop tolerating?* Everything can change tomorrow if you just stop tolerating it. As a leader on the project, or as the superintendent leading your area of the project, your standards are the ones that should and will prevail amongst the rest of the workers on-site if you lead with authority, drive, and passion. If there are behaviors that the owner, architect, or anyone else on your team would not condone, it is your duty to close the gap and correct that behavior.

You do this by making sure that everyone on-site comprehends the expectations. You make sure that you and your team are aligned with the expectations. You make sure that the contracts for your trade partners are aligned with those expectations, and that you have paid for the behaviors you're expecting. And then you systematically and relentlessly go to work to exhaust the bad behavior. That means removing bad behavior, making it impossible for the bad behavior to continue, and make it so labor-intensive to continue that behavior that people would rather fall in line with expectations. Some superintendents prefer a kinder approach. I, personally, do not have time to patiently deal with correcting behavior on a project, so I implement zero tolerance systems. But whatever you do, your duty is to make sure that the people who follow the project rules according to the owner's expectations thrive, and anyone who is not following project expectations does not thrive on the site or are removed immediately.

Construction is very similar to war. In war, bad behavior cannot be tolerated because it reduces the energy in the capacity of the army. It spreads like a cancer throughout the ranks and causes mutiny. Everyone within the army needs to understand that those who do not follow proper practices will be summarily punished or removed. This is the only way to maintain law and order among the ranks. Construction is similar, and your duty, your job, is to make sure that everyone in the ranks on your project site knows that bad behavior will be dealt with. This information needs to be made public, and people who are removed need to be told as much. If someone is let go, sent home, or terminated, a public explanation to the rest of the team is necessary. The context provided should explain that *the rest of the team was protected by removing the bad behavior.* If you do that, you will write the narrative rather than allowing the departing individual to write their own narrative and hurt the culture on the way out.

Leaders cannot tolerate bad behavior. They show this to their team by setting clear expectations and by enforcing the rules without tolerating even the least sign of dissension, bad attitudes, or disobedience. People who have bad behaviors need to be disincentivized and taught that it is easier to take care of our customers and provide remarkable results than it is to continue the bad behavior. People who need to be punished or removed from the project should be told so directly, and everyone else on the project should be told why the person was punished or removed, and everyone should know that their good behavior is being rewarded. This is how you exhaust bad behavior on a project, and there are hundreds of different ways to do so according to your own style.

Principle 4 - Mini-Marches

On a construction project, it is very important that you engage the team in mini-marches. Historically, in the military, experts found that soldiers who marched for 50 minutes followed by a 10-minute break, resulted in a 50% increase in longevity and distance in contrast to a long march with no break. Breaks provide physical and mental relief and give soldiers a sense of accomplishment. The same is true for field crews and trade partners on a project site. It is best to employ mini-marches while advancing towards milestones or to other targeted accomplishments and so provide the workforce with a sense of winning, direction, and clarity through a short interval. Some of our construction projects are so large that they can feel overwhelming and never-ending. It is very important that we have mini marches to boost morale and give us a sense of accomplishment along the way.

A leader will do this by defining and communicating anticipated targets. Leaders should reaffirm daily to every worker on the site where the project is headed and how each contributes to the whole. Construction contracts continue to incentivize companies to work in a siloed manner. Through mini-marches, we can take disconnected and siloed contracts and turn the entire workforce into a unified body of workers and professionals.

The philosophy behind mini-marches does a couple of things for us that are very important. It allows the team to feel like they are conquering. It also allows the project management to know when the opposite is true and when we need a course correction. It also increases morale by allowing shorter segments to be the focus and for us to reflect on the successes and losses of that segment. It ensures that everybody is working in the same direction. A project leader can create mini-marches by establishing the end goal while visually and verbally communicating the march, performance indicators, conditions of satisfaction, and daily progress to the entire team. This could be done with weekly visual planning, daily huddles, or a performance target that is marked on a visible graph that indicates progress to the target. Nonetheless, it must be fun and challenging as the team is rallied and unified towards the next goal. Mini-marches are one of the most important things you can do as a supervisor to invigorate your team to high-performance levels.

Principle 5 - Hobbling the Workforce

Hobbling the workforce is a military term that is mentioned in the *Art of War* by Sun Tzu. Hobbling takes place when the military general sends troops into an impossible situation without forethought and planning. The directive to go take the battle, conquer the army, cross the bridge, or defeat the enemy in an impossible situation does not lend itself to victory. Historically, military generals have acted in this way in the course of battle without thinking decisions through.

The way to overcome this strategic mistake is to make sure that we come to terms with absolute facts. We cannot simply be optimistic or trust in wishful thinking to make it through the situation with the energy of the workforce. We have to be 100% realistic with the current conditions and be able to provide and brainstorm solutions that are realistic according to the circumstances.

There must be extensive pre-planning to map out the various scenarios in which we may find our army, or in this case, our workforce. We cannot plunge our workforce into impossible situations such as meeting a false deadline. The key for a master builder is the ability to get creative, be nimble, and secure the best ideas from the team so that options can be analyzed according to circumstances.

Not very many superintendents can do this type of planning. It takes a real master builder to realize what cannot be done and to brainstorm thoroughly enough to know all of the possible options and outcomes. The recommendation for the builder is to realize the harsh realities of a situation, bring all the best people to brainstorm ideas, and find the best win-win situation to accomplish the deadline and turn the dials of manpower, cost, equipment, or method. The dials that should not be turned, or the options that should not be considered, are to have workers perform work at an unrealistic pace, undertake work for which they are not trained or prepared, and/or have workers do something that is unsafe.

The strategic advantage that master builders will employ is a firm grasp of the reality of the situation, the best ideas for available options, and the absolute commitment to finding the best solution that will conclude in a win-win for everybody involved. Most of the time we only think of one or two possible solutions when there are many. Never hobble your workforce. Do not give up until you find a solution that will work, that does not compromise your values, and that can be remarkable for the project team. If you do this you will be known as a master builder who can work with all of the resources on-site, and you will build enough trust to make it to the end of the project on time, under budget, with a quality project.

Principle 6 - Work Effectively with Your Field Engineers

A proficient superintendent will understand how to mentor field engineers. In fact, the superintendent position is incomplete without the use of field engineers on the project. If the superintendent is responsible for overseeing "how" and "when" we do things on-site, the field engineer should be tasked with arranging "where" things go on-site. It is very difficult, if not impossible, to manage this without the help of the field engineer

Field engineers are not personal assistants and they are not runners. They are responsible for layout, control, lift drawings, front-line quality and safety management. A wise superintendent will direct most engineering questions about details and drawings to the field engineer so he or she can focus on planning and executing work. In this way, a superintendent can focus on watching the project and keeping it running properly.

Another distinction between a superintendent and a field engineer is perspective. While a superintendent should ensure things are going well and that work is being done on-site in an acceptable manner, a field engineer has the task of assuming everything being built is wrong until proven otherwise. The field engineer must espouse a certain amount of productive paranoia that will force him or her to check everything and be in the details of the project. These are complementary roles that will help ensure everything is built on-site to a high level of quality.

A superintendent will do well to realize a few things. First, the field engineering position is a training position, and they will make mistakes. But those mistakes are tolerable when we consider the value added to the career of a builder when they can start in this intense and demanding position. The field engineer position is also the best way we know of to entice construction professionals to follow a field path through construction management. Second, the field engineer is a presence in the field and can be the eyes and ears of a superintendent throughout the project. They can and should ensure conformance to quality, safety, and production standards, and ensure that the directions of the superintendent are followed. Lastly, the field engineer primarily supports the work of the craft and trades. A field engineer must learn to figure things out so workers can have predictable environments on-site. Only in this way—when workers are being productive—do we make money.

A field engineer, when utilized properly, creates flow, continuity, and support for all field operations, and is a superintendent's best asset on a project.

Principle 7 - Support the Craft

Excellent superintendents make supporting the craft one of their top priorities. Ask yourself how you make money for the company. Is it when we send an email? Is it when we type a letter? Is it when we drive around? Is it during meetings?

The answer is simple--we make money when a worker is working. The transportation, fabrication, coordination, design, and management are all non-value-add actions, necessary to prepare for the moment when the worker works. The money is earned when work is being put into place. And who puts that work into place? The craft.

Craft workers are the heroes. They are the star players. They are the champions, and our industry treats them like a necessary evil, and at times, less than human. If you have found yourself operating with that mindset, I urge you to repent and rethink your life. The craft are critical to what we do! They deserve our respect and reverence.

Knowing how important the craft workers are to our work, our lives, our livelihoods, and our success, we need to optimize their work and provide them with clear instructions and training, reliable materials and information, and safe places to work. This is one of the main duties of a field engineer and a superintendent—to create stability and flow in the life of our skilled craft. Ask yourself if all the skilled craft on-site can come to work, go where they need to go, huddle with the entire project team, then go to work with the right materials, instructions, tools, and a clean work area. Is that the situation they have? Is their work fulfilling, yet uneventful? If not, we have work to do.

It sounds harsh and dramatic, but many superintendents in the industry treat the craft like slaves who need to get the job done without any support. We can change that by sharing with them what we are doing, why we are doing it, and asking their advice along the way. We can take an interest in their training and development. We can abandon patterns of wrong behavior by providing opportunities and helping the craft to meet their career goals and reach their potential. We can begin to focus on their work and give all we can to make it more effective, productive, and enjoyable.

Principle 8 - How to Use the Project Administrator

One of the most self-preserving skills a superintendent can have is the ability to partner with and use the office manager or project administrator on the project. The best generals I have ever observed always had a successful partnership with the project administrator. The range of responsibilities for a PA span from being a guardian for the project to ensuring everyone on-site has orientation and a support system. Project administrators ensure that everyone coming into the project site has the proper safety stickers or safety training indicators on their hardhats.

Keeping track of daily reports, safety inspection forms, and other typical types of reporting for the superintendent are the kinds of responsibilities that can be delegated to a project administrator. It is completely appropriate for a PA's relevance on-site to be that he or she enables the project leaders to function effectively. Anything that can be done capably by the project administrator to optimize a supervisor's time is in the best interest of the project.□

Principle 9 - How to Use the GC Carpenters

The same principles that apply to the project administrator apply to the carpenters who work on the GC/GR side of the project, though not in relation to the cost of work or as self-perform. The carpenters and laborers who assist the lead superintendent and the assistant superintendents with the project maintenance and control can be the watchdogs of the superintendents and maintain operational control. There must be real-time communication from the superintendent of the project to the on-site leaders of the laborers and foreman who work for the GC team. Each worker should follow a daily checklist. They should have clear expectations based on their work performance and their duties, and they are expected to lead, and to be the representatives of the on-site superintendents. Anything that needs to fall under their control, to create bandwidth for the superintendents, create stability for the project, and to make sure that we have a safety presence in the field should be done. Organizing your team of craft helpers on-site is one of the first things that should be done to maintain operational control.

Principle 10 – Developing Your Success

There are some points of view common with superintendents that do not create success in this industry. You would do well to self-reflect and correct behavior and attitudes if any of the following affect your performance:

- If you are too wimpy to hold people accountable.
- If you are not organized.
- If you cannot delegate.
- If you do not speak up.
- If you are not mentally alert and sharp.
- If you fail to acquire all required technical skills.
- If you communicate poorly.
- If you do not learn.
- If you are lacking drive and passion.
- If you have a bad attitude.
- If you are dishonest, meaning you don't strive to tell the truth 100% of the time.
- If you do not receive feedback with an open mind or at all.
- If you engage in criminal behavior.
- If you harass or discriminate against others.
- If you do not care about people.
- If you lack people skills.
- If you repeat mistakes.
- If you only have bad project experience.

- If you have no grit.
- If you do not give first and try to create value for the world.

Self-learning

The world of a superintendent is so complex that a complete manual would necessarily develop into a series of thick books that would fill a small building. The attempt of this work is to cover key principles that will lead a superintendent down a helpful path to find his or her own style. Below is a list of necessary skills in construction that you should commit to learning in your own organization according to company process, culture, and style. Persist in learning the following:

- Lift Drawings
- Issue Resolution
- Impact Control Planning
- Introduction to Job Cost Basics
- Self-perform Management and Production Control
- BIM Coordination
- Industry, Regional, Company, and Role Specific Legal Considerations
- IT Requirements
- Email and Calendaring Systems
- Survey Management
- Using Field Engineers

Other Items

All superintendents, by virtue of taking on the role, eventually learn about fundamentals of construction; therefore, those fundamentals will not be included in this work. Elements that you will naturally come to know and learn about include the following:

- Cranes
- Rigging and Hoisting
- Construction Methodology
- Safety
- Quality Systems
- Site Logistics Planning
- Visitor Control
- Daily Reports
- Equipment Management
- Excavations
- Structural Systems
- Project Photo Documentation
- Managing TandM Tickets
- Field Purchasing
- Weather Precautions and Management
- As-Builts
- Building Commissioning
- Shop Drawings
- Inspections and Testing

Boot Camps

My goal is to help every field builder in our industry to be successful on a daily basis, to bring respect back to individuals in construction, and to preserve the balance and well-being of families by creating remarkable work environments. If you are interested in immersive and practical training that will take you to your next level, please consider the benefits of a superintendent boot camp:

- You will have the skills to step up and hold people accountable.
- You will have the skills to avoid contention and negative project behaviors.

- You will be able to control your project with operational excellence.
- You will be a safety champion.
- You will be able to work in a disciplined way with daily habits.
- You'll be able to plan any project in pre-construction.
- You will implement lean on your project.
- You will complete all items daily that a disciplined super must do.
- You'll be an excellent builder.
- You will have advanced scheduling abilities.

Step-by-Step Guide for Superintendents:

Superintendents have to take multiple concepts of theory and put them in an order that can be easily implemented. Some prefer theoretical concepts while some prefer the simple steps of application that can be implemented one-by-one. I encourage you to use this next section by reading or listening to the steps and implementing each step on your project site.

Period 1: Prepare Yourself

Step 1: Create Your Mission, Resolutions, and Habits

Acting as a supervisor is one of one of the most strenuous roles in the construction industry--or any industry for that matter. To be successful, you must first create a mission for your personal life and for work, examine your ability and your resolve to carry out your mission, and then set up habits that will support, balance, and reinforce your effort to reach your goals.

Keep this in mind: "Success without fulfillment is the ultimate failure" - Tony Robbins, American author, coach, motivational speaker, and philanthropist.

There is a very impactful story from early in my career that stands out to me. The story depicted a distinguished leader in government and in religious circles who developed a list of resolutions early in his life. He was attentive and watchful of those with whom he interacted--his mentors and role models--and by the time he reached the age of 14, he was already aware of the mission in life that he wanted to fulfill. He chose to jot down a list of resolutions that described his chosen behaviors, what he would look like, how he would dress, and how he would conduct himself in all situations. He also outlined moral and ethical resolutions that he believed should accompany him on a successful life's journey. He kept that list on his person throughout his life and often referred back to the use of it in talks and public addresses as an anchor to his success and the foundation of who he became. I knew I had to share this story with superintendents as a key to their success as well because people with no direction typically stagnate in their career. I, too, have always kept a very clear outline of what my mission is in life, the resolutions that I must make and keep to accomplish that mission, and the habits that will support my goal.

I first learned about the concept of having a mission for my life and for work in the book, *Unbeatable Mind*. Author, Mark Divine, describes mapping out a vision of what success could look like in life--personally, with family, and at work. He challenged everyone reading the book to think about what we would want people to say about us when we are deceased, the eulogies that would be spoken at our funerals, and what we could look back on with pride. Divine challenged those of us reading the book to contemplate what we thought we were put on this Earth to learn, develop, and become, and what talents and strengths we were called to share with the rest of the world. As I reflected on the course of my life--my past failures and successes, personal development and future aspirations--I outlined my unique strengths and what I was called to share with others.

Once I put my thoughts down in writing it became clear to me what I needed to do next--develop a list of resolutions designed to help me accomplish my mission. Some of my resolutions in life are to live like nobody's watching, to give first and don't think of taking, to be kind in my approach to build people up, to use appropriate language, to always be upright and honest in my dealings, and to be transparent in everything that I do. I reference this list every day along with my mission to make sure that everything I do is achieved by a select set of resolutions that help guide my path.

I then listed certain habits or routines that I knew I would need to put into practice to help me progress because having a mission and resolutions doesn't do me much good if I don't commit to a daily habit of reading or studying those resolutions--a morning routine that reminds me of how I want to win the day. I have maintained this habit for years and consider it an attribute of my success.

As you think about this concept, ponder these few things first: Could you outline what you want your life to mean? Can you name the unique talents and abilities you have that the world needs that only you can give? Schedule your time to really contemplate these significant questions, then create your mission statement and your list of resolutions based on who you want to be. Think of your virtues, values and ethics, your role models, your mentors, or any other people in your life that you have come to respect. Then ask yourself this one question: What would you have to do consistently, weekly, monthly, or yearly to accomplish that mission and share your unique talents with the world. What habits must you develop to communicate, to share, to be remarkable, and to live a remarkable life. Write those down on whatever format you choose and refer to them daily. If you do this. you will quickly create the person that you want to be, shape how you behave, and better bring to pass the vision you have for your life.

Step 2: Begin Note-taking

Your success is ensured when correct principles are consistently and properly implemented. One of the first steps or habits to apply to yourself and to your team is the habit of note-taking. Every great builder has to discern the system that will trigger or remind them to perform specific tasks on the project site, follow up on critical items, and direct certain efforts. When triggered, these details immediately need to be written down and acted upon on a consistent and frequent basis throughout the day, week, and/or month. This will make the difference between somebody who forgets versus a builder who never forgets and is always doing the right things.

If he (or she) didn't write it down, it ain't gonna happen!

Years ago, I encountered a general superintendent who was responsible for overseeing a 150-million-dollar prison in Southern California. I was working on the project site as a new field engineer and was often asked to walk with this superintendent through the buildings where I was performing a punch list to finish the building. In our reflection walks I would often see him observe something and then utilize a small voice recorder that he kept on his person to take verbal notes. Afterwards, after many brief recordings of things to do, things to be delegated, and things to follow up on, he would return to his office and transcribe those notes onto a sheet of paper which comprised his to-do list. There was never a time that he observed something to be done when he did not write it down for follow-up and action. I decided then and there that I would find my own system by mirroring his, so I went out and bought my own voice recorder and began immediately creating a to-do list. Over the course of the next few weeks and months, everything I thought needed to be done, everything that triggered my impulses to make improvements, and anything that somebody delegated to me or told me to do, I would note it on my voice recorder then write it on my list. I checked my list throughout the day and followed up on items. This habit has made a tremendous difference for me and ensures that I never forget my duties and responsibilities.

Any time a senior builder or mentor is walking the project site with a fellow worker and a moment comes up where there is an opportunity to delegate an item, the first thing that mentor will do is look to see if that person is already following up by writing it down. There is absolutely no way a person can remember all of the things that are to be done. There is absolutely no way a person with a good memory can remember to follow up at the exact time and with the right context. We often hear the following in the construction industry: "I have a good memory. I don't need to write it down." These are the most unorganized people because I can absolutely guarantee that they will neglect key items. There is nobody so intelligent, so sharp, or with a good enough memory that they do not need to write down to-do items. Additionally, if somebody tries to remember specific items, those things are being stored in their active memory, causing stress, and reducing the capacity to be mentally present or remember additional items that are needed in the moment. The bottom line is this: if you don't make a note of it, more than likely it isn't going to get done. And people like that do not succeed in the construction industry because they always forget.

I often tell new field engineers and superintendents that they can never fail to remember what needs to be done. Whether it is to be delegated, remembered, or simply observed as something to be executed in the field or office, it cannot be forgotten. People who cannot remember their assignments and don't follow up in an appropriate time frame are hazardous to the remainder of the team because they burden others with their responsibilities. Conversely, if an employee will remember those things and follow-up in a timely manner, they will be fulfilled and successful and will always receive the additional assignments and promotions because they can be counted on to execute their portion of their work within their role.

As a first step, I recommend that you explore and experiment to find a system that works for you. This can be on computer, paper, or voice recorder. Whatever works for you, the key is to get started and to get started now. Try this: For at least 60 days write down everything that you think needs to be done, everything that is delegated to you, and everything you are told to do. Reference that list at least three times throughout the day and prioritize the things that need to be done early in the morning. Continue to work on your system, adjust as needed, and frequently reinvent yourself to find quicker and more effective methods. If you will follow this practice for 60 days, you will form a habit that will serve you throughout your life.

You will be known as an employee who does not need to be managed in detail, as somebody who always follows up, and as somebody who always remembers. You will be known as an employee that stays within his or her role, always executes, and can be trusted and therefore valued as a true team member. Promotions will come easier, your performance will be consistent, and you will constantly improve. You will be given more important and more strenuous responsibilities throughout your career. As you develop this habit, you will begin to notice more and more things that are crucial to the success of your project. Your subconscious mind will begin to work on problems in quiet times throughout the day and provide solutions for you. You will write those things down, act on them, and always seem to know what needs to be done at just the right time. You will be known as somebody who can plan and see into the future, somebody who is always anticipating, and somebody who is always prepared. Those who are watching will say that you are organized and always on top of things, and you will not have to deal with the failure of poor performance, forgetfulness, or lack of execution.

Step 3: Anticipate your Next Project Assignment

On construction projects, the way we tailor our role, strategize, and employ tactics is entirely dependent on the type of project we will be working on. When possible, anticipate what your next project assignment will be. What is the type of building? What is the square footage? Where will this be geographically? Who composes the team? What will your role be on the team, and what scopes or geographical areas will you oversee? Are there any specialty tasks to which you will be assigned? If you can anticipate these things, you will be able to properly identify your role and form a plan for success. I encourage you to talk to team members, visit the project, study any preliminary drawings, and understand and outline your role before moving forward.

Step 4: Know Thyself

It is very important that you know who you are before you become a member of a team. Now, that might sound a little bit silly because you think you already know yourself, but there are specific strengths, weaknesses, and tendencies you may take for granted that will come to bear when connecting with your team. It is important that you know your strengths and weaknesses, your style of communication, and the best way you have of providing insight to the rest of the team on how interaction with you can be done in a remarkable way. There are many different ways to self-discover and communicate who you are. We will talk about a few of these in this section.

"Know Thyself." - Tony Robbins, American author, coach, motivational speaker, and philanthropist.

Great teams are built based upon trust, team accountability, setting goals together towards a common vision, holding each other accountable, and performing. Some would say that personalities don't have a place in the team, but personalities greatly influence how we engage in all five steps of team building. Most specifically and importantly, how we build trust with each other. If there is a step before building trust, it is to know one another, but for others to know you, you must first know yourself. In a team, you and the others must know your skills, your style, your strengths, and weaknesses. They must know what triggers your reactions, what traits need to be called out or corrected, and your team must know your preferred method of communication when you reach the point of stress. They must know your hot spots or your trigger points. All of these characteristics combine and form a picture that informs team members they can build trust with you, communicate with you, and resolve issues so together you can collectively head towards performance.

Some of the best self-reflection comes from doing questionnaires. These questionnaires can help gain insight due to the variety of responses. Some of these questions are as follows:

- In one word, what do you need from your team?
- When you have reached a point of stress, what is the best way for others to communicate with you?
- What is your biggest strength?
- What is your biggest weakness?
- How do you like to be called out?
- What is one thing your team should know?
- What is your preferred method of communication?

Another way to fully understand yourself would be to do the appropriate personality profiles. There are many in the industry, but the main profiles in the construction industry that have historically been very helpful are the Myers-Briggs Type Indicator, the StrengthsFinder profile, and the determination of your learning style, be it visual, auditory, or kinesthetic. Most of these tests can be done for free online. The best practice that we have found, is to take these tests and distill down the most important and pertinent points into what is called a player card for your team to use and reference during team interactions. This player card, complete with the most important bullet points that describe your personality, can be discussed weekly in team meetings, posted at your desk, but most importantly when filled out, will help you to gain an insight into who you really are.

Now, a word of caution... some less favorable, but deeply embedded personality traits will be very difficult, but not impossible, to change. I recommend that you ask, "how can I turn my weaknesses into neutrals, and how can I turn my strengths into game-changing performance?" Most importantly, if your team knows how to communicate with you, and you know how you can use your natural abilities and strengths to communicate with your team, then you can build safe pathways to communication, and then the team has the ability to learn and build trust and begin a very difficult path to accountability and engaging in healthy conflict.

If you concentrate on positive self-improvement, you will stop wasting time trying to improve your weaknesses. You will spend more time flexing to neutralize your weaknesses, and you will put your strengths to work in a way that will bring you more success, fulfillment, and money. If you share who you are openly and naturally you will build rapport with other people. They will see you for who you truly are, and have a softer touch or softer feelings towards you because they know where you're coming from. Others will see commonalities and traits that they appreciate and will begin to realize how your unique abilities fit within the team. You must focus on what will bring you a return in your success which will help you to stop wasting your time. If you do what you say you will do, if people know who you are, and if you actively try to communicate and build trust, you will be on your way to being a remarkable team player as long as you combine the associated traits of being humble, hungry, and smart.

Step 5: Outline Your Style

In addition to knowing your personality type and your tendencies and preferences, it is also crucial that you understand your style. The misalignment of individual styles on a project can cause a lot of contention within the team and especially with others with whom you work. Your style is your unique approach to how you accomplish things with management and leadership. It is how you think about things, see things, and approach things. Knowing your unique style will help you in all of your interactions, help you understand why you do what you do, and help others to predict certain actions and know where you're coming from.

Everyone has a style with which they approach their work.

I remember being an assistant superintendent under one of the best project superintendents I've ever worked with. At the beginning of the project he made clear to me, and the other assistant superintendent and field engineers, exactly how he wanted to run the project. He taught us how he liked to run the schedule, treat trade partners, and operate the site. There wasn't anything in his style that we didn't clearly understand. Knowing this, we were able to align behind his leadership, anticipate his actions and decisions, and be one united force in the field. When trade partners would ask him questions, we always knew the common vision, the style of the lead superintendent, and were able to give united answers. This showed me the importance of communicating styles and having a united support for that style with the project leader.

The leadership style of people on the team is very important. You may have people who are somewhat hasty while others may be cautious. You might have people who stylistically are more politically correct and you may have others who are the polar opposite. You might have people who are thoughtful and cautious and treat everybody equally while their counterparts are judgmental, decisive, and tend to play favorites. These differences in no way mean that people cannot work effectively with one another if they are known to the team. The unifying component for project managers and supervisors is that they align in common ground found in their approach to the project and lead by example in a style that assistant superintendents and field engineers can adopt as they execute the project.

Familiarity with leadership styles means team members are able to function properly in appropriate situations. During World War II, General Eisenhower's style was better for his position as the Supreme Allied Commander. He was able to unify the Allied Forces through political acumen, balanced leadership, and fair treatment of the other allies. General Patton, on the other hand, was very decisive, judgmental, and not very politically correct. Patton drove hard, trained hard, and treated the enemy with a fanatical sense of contempt. He was known to grandstand, he led from the front, and was a genius at military tactics. Patton was instrumental on the front lines during World War II, and General Eisenhower was needed at command. Each person who leads has a unique style that suits their position on their team. I have identified my unique leadership style as outlined below. As you can see, the ways I implement my leadership is very clear.

Jason's Style:

- Win the war before going to battle.
- Have smaller, more nimble units, and have an unlimited amount of options to adapt and react.
- Stabilize everything. Success only in stable environments.
- Have a zero-tolerance policy for safety, cleanliness, and organization.
- Build the team first.
- Make, "respect for people," the basis of all decisions.
- Transparency in all things.
- Always do the right thing.

Identifying your style is not easy. You have to intentionally think of how you do things. Constructive thinking takes discipline. Use an appropriate amount of time to ponder and make notes on boards that are concise and meaningful. Write your outcomes down and make it a part of your player card. Communicate your style preferences to your team, and make sure to identify differences with other members of your team so they can be reconciled. Always refer back to your defined leadership intentions and use them to identify updates or improvements in your style. Look at ways that may not be in keeping with your future goals. Adjust as necessary.

If you can identify your style of leadership, properly communicate it, and align with other members of your team, you can leverage an incredible amount of success collectively. You can lead with unity alongside your other leaders, and also know which leadership styles are best suited for certain activities. You will be better utilized in certain situations by others on your team because they know how you adapt and react. Knowing these styles, will fully engrain these characteristics into your leadership, and define your success in the future. It will galvanize how you lead and who you are on the battlefield.

Step 6: Develop **Your Role**

Once you know your style of leadership and you know what your assignment is, you need to develop your role on the project. It is important that you know the things that you should be doing to support your team and the things that you are expected to do. These things should be focused, effective, and get to the heart of the issue.

Teams work best when everyone has clearly defined roles. Where the lines of definition are thin, others can lean in and overlap to support each other cross-functionally.

I remember a time when I had a clearly defined role in construction as a lead superintendent. I identified my role as creating the plan, seeing the future, allocating manpower, ensuring procurement was on-time, and making sure that I planned work so that assistant superintendents could plan and execute the work in the field on time and under budget. I wrote these down and communicated my listings to everyone else on-site. My actions were very effective on the project because people could count on me to do certain things. If the project manager or an assistant superintendent was gone, I would cover for them in their role. Likewise, the other people on the team would cover for me in my role when I was gone, on vacation, or tied up with a pressing matter. Everything on-site was standardized in a stable environment so people could lean in and perform my role at a moment's notice. We always did cross functional training and everyone was able to lean in when necessary.

If you have a cross-functionally trained team, with a horizontal feel, where people can lean into each other's roles, even though they have clearly defined roles and expectations, then you will never have a gap on your project team that cannot be filled. If the project manager is gone, anyone can go in and lead that meeting. If the project superintendent is gone, anyone can lead the trade partner weekly tactical meeting. It will make for an environment where your owner or the customers will always feel confident that you will accomplish the things necessary on your project. There should be an intentional process to cross-train, opportunities for other people to flex in and perform certain parts of your role, and there should be the ability to switch who runs meetings so that the meeting systems can always be maintained, regardless of who is on-site at any given time.

To implement a smooth flow of cross-training, create a list describing your role on-site according to your skills and according to the project needs. Make your list available to the entire project team. The list should not be superfluous, but rather meaningful in its impact and description. Descriptions for a senior superintendent would explain duties such as ensuring proper procurement, planning and executing work, maintaining a balance for manpower on-site, seeing the future, and always taking care of neighbors. There should never be any redundant duties like overseeing another role, or filling in certain reports or forms, or attending meetings. Everything listed in your role should be meaningful and impactful. Create an intentional way for people inside the organization to do cross-functional training, and have different people take turns running meetings.

If you do this you will always have proper coverage, people will meet their career goals, and there will be very little duplication of efforts among the team. You will be able to have smaller teams which will be more effective because they are 100% utilized to their capacity. You will have a team that will be able to dynamically flex into each other's roles without the need to do or duplicate someone else's role when there's no need. You will have a team that is well-trained to do anything and accomplish any task at any time.

Step 7: Develop Your Leader Standard Work

Leader standard work is the standard work in a leader's day that has to be accomplished in order for him or her to be effective. Twenty percent of what we do will bring 80% of our returns at work. Leader standard work enables your 20% and ensures that we are focused on the right things and avoiding chaos. One of the first things that a leader will do in developing his or her leadership strategy is to develop leader standard work that will support the role that was just developed in the previous section.

Twenty percent of what we do brings 80% of our returns. Leader standard work enables your 20%.

People have often said to me, "You do the work of two or three people, Jason." I've heard this comment over and over and over half of my career. I attribute this to the lean concept of creating leaders standard work. I learned this in a Leancor eight-week training course. The concept is that we need to identify the things that we should do on a standard basis that will make us more effective. I did that on one of the projects where I was the lead superintendent and it made all the difference. My time in my role was so effective, I was able to leave the project for extended periods of time to work on proposals and other assignments, I was never missed, and the project was always operationally stable.

The number one thing that should be on your leader's standard work is your family and personal time. Things like date night, baseball games, other training, workout exercises, and similar activities should be put on your personal calendar. The second thing should be your leader's standard work--which is the time set aside for your work assignments dedicated to the effectiveness of your role and based on the standard meeting system of the project so as to protect against chaos in your personal schedule. Other meetings, impromptu requests, job walks, any questions from the trade partners, etc., would be done around your leader's standard work. Even if there was a high-profile person on-site, the goal is to always protect your leader's standard work so that you can be effective in your role. People that do not do this will usually be a victim of chaos on-site, the victim to their emails and their messaging systems, and be subservient to somebody else's schedule.

To implement your goals, use an Excel sheet or start working on a calendar to time block certain things that are important for your leader's standard work. You do want to include some blank space for occasional chaos and other developments that may occur, but for the most part, you will want to put in your personal and family items and then fill up 40% of the remaining calendar space with your leader's standard work. This should be done around your team structure in the meetings that you attend. You will align this with your role that you created and make sure that it is impactful. The question is this: Do you feel like your leader's standard work, if followed, will make for a remarkable execution of your role? If the answer is no, then continue to revise it. Put this in a visible place, whether it is your calendar, or whether it is an Excel sheet that you post on the wall. Make sure people know that you will protect the standard work so that you can fulfill your obligations within your team.

The promise for this principle is that people will notice and say that you do the work of 2 or 3 people because you will get your work done on time, the project will be able to run without you, and you will be able to spend enough time at home to keep a work life balance. You will always be ahead of the game, ahead of activities that you're in charge of, and also have a great return in current growth and promotions.

Step 8: Develop Your Personal Organization System

Every great builder has a personal organization system. Most great builders learn this when they're in the craft role, come up as a foreman, or learn as a field engineer. Having a written to-do list that is referenced more than one time throughout the day is one of the single most important things for a master builder. There are literally millions of things on a project that need to be done. Knowing which ones take priority is one of the main skills of a master builder.

No one on this earth can remember what they need to remember in any role without writing it down.

Remember the general superintendent who used a voice recorder to compose his to-do list? He would take that information to write new lists which he delegated to other people on the project team. His lists showed what winning looks like and triggered everybody to be on schedule ahead of the game and effective.

No one can remember what they need to remember

without a system. Each person on a project site needs to be able to systematically work through their tasks on a daily basis. This comes from being able to take inventory of your to-do list, highlight them in order of importance, then do what needs to be done, when it needs to be done. I cannot emphasize this enough. Being effective is driven by your personal organization system. Continue to explore and experiment to find a system of order that works seamlessly with your personality type.

Everything in your career will benefit from this system and by good habits of referencing your notes on a daily, even hourly basis. People that have personal organization systems are effective, disciplined, meet their obligations, show up to meetings on time, and always return their assignments in the appropriate amount of time. People will come to understand that you are somebody upon whom the team can rely. You will be given promotions, be recognized for your success, and you will be able to learn and grow more quickly than somebody who does not have the system.

Step 9: Develop Your Personal Training Plan

Every master builder must continually receive training in order to be effective in their role. This training must be intentional. Our industry offers exceptional instruction that includes safety, lean, integrated project delivery, construction systems, concrete formwork, and many other topics. A person who comes into construction and only learns through observation runs the risk of being only as effective as the individuals he or she observes. Correct principles, new techniques, and the leading industry practices can be found in industry training, YouTube, and live events and forums that will sharpen your mind to make you more successful in your career.

Your training will be the biggest single strategic advantage in your career. People will pay you according to the amount of training you have.

The general contractors at Hensel Phelps are widely known throughout the industry for their training expertise. One of the sayings within the company is that anyone who works there for five years can write their own ticket for whatever career they decide to pursue. I remember the director of training commenting on the amount of hours and new classes that were available within the company. I didn't understand the strategic value of training then and initially wondered if it was worth my time. As the years have passed, I notice that anyone who comes from Hensel Phelps has the foundation to be a master builder and are always adept at construction process, safety, and other general building techniques. Any builder who comes from Hensel Phelps will be effective as a builder because they have a solid foundation of training which is used as a strategic initiative for their own people.

The outcome of stagnation is never success. Builders need to constantly keep up to date with new technologies, new processes, philosophies like lean, and new ways of partnering with design teams and owners such as integrated project delivery. The builders who are constantly being trained to come and constantly sharpen their saws (as Stephen Covey would say) are the ones who will have the highest net income in our industry. I highly recommend that you develop a plan to financially invest in training and arrange training intentionally within your schedule. Align your training goals in accordance with where you want to go and who you want to become, and then outline when and where you will train throughout the year. Budget appropriately and work closely with your employer to make sure you are able to train. When you've completed training be sure to implement the things that you learned and make sure that you teach the same concepts to others in the industry. There's no better way to retain the information than to teach the same model.

Implement steps to identify how additional education will support you in your role--especially for you to receive that next promotion. Research the specific training that will benefit you the most, who provides it, their location, and cost. Form a training plan with your employer and record it on your yearly schedule. Be sure to attend these training sessions and come back and implement what you learn. Do this every year from now until the end of your career and you will reap financial returns that you would not have otherwise expected.

Make training one of the main focuses within your career on an annual basis, and it is very probable you will receive promotions at an accelerated rate. You will be viewed as one of the top performing superintendents in your company, and very likely within your local geographical area. You will be at the top of your class of superintendents and your advice will be highly sought. Chances are overwhelming that you will make more money and go further and faster in your career than you otherwise would. Always take time to invest in yourself because your mind is your most important asset.

Step 10: Begin Learning Lean as a Primary Focus

Lean, in construction, is a trend, a business management system, and a philosophy that is not going away. Construction has benefited by this philosophy to the extent that projects are being delivered on time, under budget, with remarkable quality, and with team health at a higher rate than ever before. It is best for a superintendent to begin their lean journey immediately because it can take years to learn enough to make a difference in your specific role according to your style.

Lean is a management philosophy, a way of thinking, and a model that will improve every single aspect of your work and life.

Lean became my philosophy after reading the book *2 Second Lean* by Paul Akers. The knowledge I gained had an immediate and positive effect on my role as a superintendent. It was easier to manage trades, remove waste, plan work, and increase stable systems. Concepts like *just in time delivery, bringing problems to the surface*, and, *morning huddles* have made a difference. So much so that I cannot imagine how we supervised work before we understood these concepts. Lean is not something that we have added to our management system, it has *become* our management system. We are now twice as effective as we used to be with the added ability to alter the course of our entire industry.

It is unfortunate that there is not a single book or training that will allow us to quickly learn about lean concepts and philosophies. It is a journey of trial and error, research and development, and will take experience and time to learn the concepts that will eventually become part of who you are and how you think. You must learn from the best books and trainings as fast as you possibly can, and to boldly and bravely implement every step along the way. Construction is heading in the lean direction and a superintendent will not be able to survive, or thrive, in the construction industry without understanding lean principles and techniques.

Expand your reading library by acquiring the best books recommended by your peers and company. I advocate the following books as a start:

- 2 Second Lean
- The Toyota Way
- The Goal
- The Lean Builder
- This is Lean

Begin studying and implementing these books, and most importantly, share as you go. One of the reasons that I recommend reading *2 Second Lean* at the onset of your journey is because it also recommends one of the best ways known in this industry to scale ideas and to share. Paul Akers and the content that he delivers is uniquely suited to spread this message far and wide.

Every superintendent who learns in-depth lean principles is more effective in their role and runs an operationally excellent project. Every superintendent who learns lean principles can better align with trade partners and vendors who are already taking their own lean journeys. It's never too soon to begin learning lean principles.

Step 11: Dress the Part

Proper dress and grooming will have a positive effect on the way others see you. Dress for your role as a professional to instill confidence and competence.

To perform in your role, you have to look your role, act your role, and appear to understand your role.

When I worked as a field operations director, I remember speaking with a general superintendent about an up-and-coming concrete superintendent, the challenges he was facing in his role and the current behaviors he was choosing. At the end of the conversation the general superintendent said to me, "Jason, all you have to do is look at how he dresses, how he presents himself, and how he keeps his desk. That's all I've got to say. That tells the entire story. He needs help or he isn't going to do any better."

There is one very true principle that we cannot argue our way out of and that is this: How we do one thing is how we do everything. Appearance matters and neatness counts. Whether it's a tucked or untucked shirt, combed hair, an organized desk, or clean truck--appearance indicates leadership and is a key indicator of how someone carries out his or her supervisory duties. If you have ever seen the movie *Patton*, you may recall that the military general was always well groomed, always neatly dressed, always had numerous pistols on his person, always ready for battle, ready to go to war. His bearing was clean and orderly and gave credence to how he executed his orders as a war general. As a supervisor, you will not exude confidence, receive the promotions you feel you deserve, or be self-disciplined in any other area of your life until you first master your grooming and your appearance.

To implement this principle, ask a trusted friend to help you design your personal appearance. How you want people to view you and your performance in your role should be intentional. Decide on the standards you will use for grooming in appearance. Pick a wardrobe that is becoming of your position. Make sure that your vehicle, your desk, and any other work areas for you are neat and orderly. Make sure they are designed and communicate the message that you are proficient in your field.

It is highly likely you will receive quicker promotions and have more self-discipline and confidence in everything you do. You'll be recognized as somebody who does everything within their power to be excellent and precise. You will be highly thought of, sought after, and you will be an example to everyone else coming up in this industry.

Step 12: Begin the Habit of Studying Your Drawings and Trigger Action

Every master builder studies their drawings on a daily basis. The minimum amount of time recommended for this is 30 minutes per day. There are superintendents in our industry who don't make it a habit to study the drawings on a daily basis, but they are brokers, not builders.

Builders study drawings. Brokers do not.

I remember one of my first roles as an assistant superintendent--and actually also in my role as a lead superintendent on the next project--feeling very strongly that I needed to study the drawings for 30 minutes every day. This habit ended up saving me from many possible mistakes, schedule delays, and problems. I was aware of what we were supposed to build within the next 6 to 12 weeks and was able to trigger things like the completion of submittals, the preparation of work, and the coordination of certain details.

Everything we know to do, everything we anticipate, and everything we delegate comes from a source. Reading the drawings is one of the main ways a superintendent can be an effective leader. Reading drawings can be a trigger to anticipate upcoming work, how to prepare for it, and forecast details needed for coordination as well as the requests for manpower, materials, and information. We must remember that the superintendent's main role is to see the future and he or she can only see the future if they have a clear vision of what the building is supposed to look like.

Implement the morning habit of setting aside time to study the drawings for 30 minutes. Whether these are electronically studied or printed out does not matter. Even if the set of drawings isn't posted with the current RFIs, it will still give you a general perspective of how the building is supposed to go together. While studying the drawings, delegate assignments to other team members. Ask your project engineer to prepare preparatory meetings for key scopes. Request trade partners to look at certain details and to be ready for materials. Ask the project manager to prepare for the buyout of certain scopes coming up. Your habit of studying the drawings will make a considerable difference in your ability to be a builder more than almost any other habit. People who know the building through the drawings are always anticipating, preparing, preventing, and being effective with projects on time and under budget.

Step 13: Get Your Tools and Equipment

Getting your tools and equipment might seem like an obvious step in the course of your work, but it is not. Making sure that you have everything you need for your role as a superintendent from a harness to your hardhat light to the right gloves, should all be intentionally designed for the actions you take in your role.

For anyone to be successful, he or she needs the tools and equipment, the right training, a place to work, and the time to perform in the role."

As a field engineer on a large prison project, our team had a set of Kawasaki mules. On the back of one of the mules we built a plywood toolbox which had a plan table, places for our tripods, a foam pad, and a box for our total station. Everywhere we went on our project we always had all of the right tools and equipment at our disposal and we were 100% effective. That organization quickly transitioned to being inside the building with a punchlist cart, a mobile desk, and a rolling chair for punch list. The awareness to have my tools conveniently at hand continued into my role as a superintendent on a research laboratory where the tools that I bought were more specifically the GoPro needed for certain inspections, the hardhat lights for doing electrical upgrades, and the elevated desks with plan table chairs that were utilized to quickly move in and out of the office without delay or distraction. This hands-on experience has always been an anchor in my career.

As I stated earlier, tools and equipment are not costly items in the grand scheme of a project. The majority of cost is closely tied to wasted man hours. That is why I have always been an advocate for buying a suitable supply of computer screens and appropriate desk space, obtaining relevant software, and purchasing the right tools that will enable us to go do our work in the field--whether it be a field desk, field equipment, or tools to help us with inspections such as drones, video sticks, or video equipment. We must have the right tools and equipment for the project. Before any project begins, I strongly recommend researching necessary equipment, then making those purchases so you can be at full capacity on the project. You will set an example for others, will not be incapacitated by a lack of a tool, nor will you be surprised when you don't have the right tool for the job.

Step 14: Get Addicted to Cleanliness and Organization

All of the great builders in construction are clean and organized. It is a mental state and a fundamental belief system. Cleanliness and organization will affect every other single thing that you do in your career whether it be planning, the execution of construction, or closing out tasks.

Cleanliness is leadership

I remember traveling the United States working for Hensel Phelps and feeling privileged to observe some of the biggest, brightest and best general superintendents in the industry. Every single one of them would talk about trying to increase cleanliness and organization as one of their first and top priorities. It was almost as if cleanliness and organization meant safety and respect and drove every other aspect of the project. It was only years later when I was asked by a project manager to be cleaner and more organized in my duties that the big picture came into focus. Cleanliness and organization is the driver for everything else on-site. A clean mind is an intelligent mind. A clean job is a good job. It has made the difference in my career.

Anyone who is not clean and organized will suffer and struggle throughout their career. They will find it difficult to control the behaviors of people on the project as well as provide successful environments for trade partners.

Cleanliness and organization has a domino effect that enables all other successful habits. If a superintendent does not practice cleanliness and organization, he or she will fail at numerous other activities, settle into a set of low expectations, and limit their career progress and effectiveness.

Begin experimenting with cleanliness immediately by controlling worksites or areas where you supervise your work. Finding systems to get this done will only improve the workers and the crews, but it will also improve your effectiveness and productivity on the jobsite. Raise your mental setpoint and your expectations. Find a small win and keep going. Never settle and never stop your pursuit to keep things clean and organized on the project.

If you do this, your owners will always comment on your well-run project. You will build trust with the Inspectors who come to your project site and see how you care for the project. Your ability to get trade partners and workers to keep areas clean will enable you to get them to do other things which are crucial to the operations of the project. Your supervisors will always know that you run a clean and safe project and you will be ready for the next promotion faster than your contemporaries.

Step 15: Drive with Urgency

A superintendent without a sense of urgency is ineffective to the point of uselessness. Superintendents must keep pace with the project to finish on time. Superintendents are the timekeepers, the pushers, and the ones who keep everyone accountable to follow a certain rhythm on the project. Their role is to pull all aspects of the project into the dimension of time and without that crucial awareness, a supervisor cannot excel.

It is the superintendent's job to bring everything on the project into the aspect of time.

As a director coaching a team, I remember when the team was six weeks behind schedule and were finishing concrete. The field director told us we needed to begin a swing shift for concrete crews and rodbusters to pick up the time the very next day. "Do you have any response from the team why they couldn't do this?" the field director asked. "If we decide now and implement tomorrow it will be done, then we will be on our way to picking up the rest of the time." After this suggestion, the team still did not exhibit the urgency to drive the situation but rather focused more on their fears and apprehensions. On the field director's next visit he had some choice, but appropriate words for the project team when he said, "We're doing this tomorrow! What do you guys not understand?" As they started giving excuses, the local lead superintendent committed to get the ball rolling and within 24 hours had rodbusters and a crew of carpenters to come in at night with the appropriate rigging and equipment. The project projections immediately went from six weeks behind to two weeks behind with the prospect of also getting back the additional two weeks.

Success in warfare has never been associated with long waiting. The quote from General Patton is appropriate for this topic--" A good plan violently executed today is better than a perfect plan executed next week." We must have a sense of urgency at work. There are things that we overthink, situations that we overanalyze, and plans that we try to over perfect in construction. This means that we frequently lose valuable time, losing days to save hours. Superintendents who have not found a sense of urgency and do not feel the responsibility to finish on time hinder the project in incalculable ways. These supervisors are among those who fail on their performance reviews, get stuck in their role, and are not ready for the next promotion.

Immerse yourself in your schedule for at least 30 minutes every day. Everything brought to your attention by your trade partners and/or others in meetings should be checked against your schedule to ensure you have the indispensable sense of urgency to get the work done on time. Also, practice coming out of the ground quickly on projects when you only have a few trades on-site. Don't overthink and overanalyze unless it is very high risk. Get the consensus of the group once you have a good enough plan and make sure you move forward with drive and urgency. Intentionally practice this on a continual basis. Become creative on how you can gain ground and move things forward as long as it doesn't interrupt the flow. For every question on-site, ask if the situation is being given the appropriate amount of urgency. Most importantly, when a problem arises on-site, do not delay fixing it. Do what needs to be done--remove that person, add that other person to the conversation, order those materials, demolish the defective work--do whatever needs to be done and do not wait. Expediency is associated with success in warfare and in construction.

If you can properly learn how to instill a daily sense of urgency, and passion and drive in your role, you will obtain success because you will not be hindered by the consequences of delay. You will have a balanced life and be effective as a supervisor because things are not piling up and burning you out. You will finish projects earlier because you authoritatively and intelligently take advantage of certain situations which do not bring the project out to flow. You will gain time and contingencies on the project and it will serve you later when mistakes arrive. You will be able to overcome and adapt.

Step 16: Lead with Drive and Passion

Every committed superintendent in the construction industry has a passion for what they do. You will hear them say, "I just love what I do," "I love building buildings," and "I love coming to work every day." Passion and enthusiasm for what we do is one of the single most important leadership qualities and the secret sauce for what it takes to build and drive a team.

Passion is POWERFUL.... Nothing was ever achieved without it and nothing can take its place.

I remember the first time that I was part of a large mat footing placement on a prison project in Victorville California as a new field engineer. The sound of the pumps, the smell of the concrete, the chill in the air, and the sound of the light plants buzzing in the morning night added to the look and feel of construction which I came to love. I remember being in my personal protective equipment and heading out to do an inspection ahead of the concrete placement. We were waiting for concrete, everything was prepared, and the crews were poised and ready to go. There was a pause in which I was able to wait and just observe the scenery. It was at that moment that I knew that I had developed an innate and profound love of construction.

We cannot look at construction as a 9-to-5 job we are simply paid to do. Construction has to be a passion, it has to be a commitment, and we have to be engaged. The best builders in this industry love what they are doing and would sometimes even consider doing it for free. They chose this industry because they love it, and because they love it they do a good job. Hopefully they have a family at home that supports their career choice and the long hours they put in for a job well done. For the supervisor, a feeling of excitement and not dread must greet the morning. This is the attitude that has to permeate our minds for a winning outcome. We need to find our passion and let it drive us then share that same enthusiasm with the families who wait for our return.

Submerge yourself into situations that you love. Notice the hotspots or the bright spots as you go about your work. If you can't find enthusiasm in what you are doing, I urge you to seriously consider doing something else. Construction is too intense, too stressful, and too time-consuming to do it simply to earn a paycheck. People who eagerly run towards their work always do a better job and often feel more fulfilled and self-assured. If you are running away from your work because there are too many aspects of it that you do not like, you will not be fulfilled and therefore, will not feel successful. Don't live anything less than a remarkable life. Do not waste your time. If you have a passion and you love what you do, there will be joy in your life. Your personal and work lives will blend naturally and that balance will lead to less stress and more happiness.

Step 17: Take Care of Your Health

It is a well-known fact that superintendents experience job related stress and may unwittingly sacrifice their health in the workplace.

We work to live; we do not live to work.

So many times on project sites there are unhealthy snacks, lengthy work hours, and a lack of focus on providing a healthy environment for the individual as well as the team. That mindset changed for me on one project in which the team held a laser focus on what it would take to be healthy on the project. Health was a part of our team meetings, healthy snacks were provided, and a group began jogging together on a weekly basis. We encouraged each other to make and keep doctor's appointments to make sure we were taking care of our mental and spiritual welfare. Even though it sounds a little unorthodox, I remember that the local university masseuse who served the occupants of the building would stop by our office to inquire if anyone needed a massage. She would set up a table and provide on-site massage therapy for the workers and management team to help relieve stress and relax. Another bonus in the main office included a complete workout room that included showers. There were also weekly yoga classes, benefits for counseling as well as benefits for preventative care. These extras were advantageous to help maintain balance at work. What I learned through this experience was that the better we take care of our people--investing in their minds and bodies--the better off they will be coming to work sharp, healthy, and refreshed. Our bodies are the conduit through which our minds are used and are activated.

Superintendents should not be part of the demographic who have heart attacks at the age of 36, but this is a common occurrence in our industry. In fact, it is not unusual for television commercials to target the construction industry with ads for things like testosterone therapy due to stress. It is because our industry allows us to fall victim to the schedule of chaos. We have leaders and workers who do not take care of their minds and bodies and we absolutely need to make our health a priority.

Utilize your calendar to include your dental appointments, doctors' appointments, and follow up appointments. Put your personal health items onto your leader's standard work any time that you personally need for meditation, exercise, or other mental and physical health issues. Do not let project circumstances derail you from partaking in self-care. Use your PTO, health insurance, dental insurance, and/or counseling services. Get over the stigma of working out before or after work, having a masseuse visit the job site, or having healthy snacks inside the trailer. Create an intentional healthy culture on the project.

There is a way to do what we do and to also be healthy. There is a way to do what we are doing and to get enough sleep. There is a way to do what we do and to have the right coverage on your project so we can attend our appointments and retain our health. There is a way for us to do those things which would allow our bodies and minds to be at full capacity. We will be more successful, more fulfilled, and happier. We will live a longer and more productive life if we take care of our bodies throughout construction. As more of us set an example for others to follow we can change the lives of hundreds, even thousands of people.

Step 18: Identify your Moral and Ethical Standards

Being ethical and being moral cannot be overlooked in the field of construction. These may be human traits that people think should be left to preachers and the teachings of parents, but are an unequivocal part of your character--especially if you are to be successful at work.

Morals and ethics create the foundation. Without them, all other success will fail.

When I was a new lead field engineer on a project in Austin, Texas, I found myself experiencing substantial difficulty because I was not honest, I did not have a moral code, and I did not do ethical things. I existed in that state of mind for three long years but found my way out with the discovery of some religious texts that introduced the aspect of leadership into my life. Everything for me began to change after that. Although the studying that I did had nothing to do with work, and appropriately remained outside of work, it affected who I was and how I showed up at work.

Everyone on-site needs a code of ethics and a sense of morality in order to make ethical and moral decisions on the project site. From that standpoint, who we are is more important than what we think we ought to do. We need a foundation built on doing the right thing and personal integrity in our interactions with others. If we do not have that firm foundation, we have no defense should we ever be accused of being dishonest, unethical, or not following the standards of the company.

Satisfying the paradigm of morals and ethics can be difficult. I recommend that you return to your roots and give careful scrutiny to your life from a moral and ethical perspective. Seek out self-help books. If you have a spiritual or a religious affiliation, the best thing you can do is to continue those habits to strengthen your dedication. If you realize there's a deficit in your ethics and morals, educate yourself on ways to improve. When asked, I often say the best thing that a person can do to be successful in construction is to become very familiar with their personal beliefs and the personal dedication to their religion or their spiritual well-being.

So many times in construction people have to be let go, disciplined, or demoted because they do not have a moral or ethical background. It is quite sad to explain to somebody why these consequences are coming to pass when they don't fully understand it. Remain sharp, and if you do so, and if you have the foundation, you will never be caught by surprise with this. You will be able to make moral and ethical decisions, and you will be able to always know and do the right thing on the project site, and therefore, make better decisions. You will be looked up to as a person of character, and you will be ready to change the lives of many people on your project sites.

Step 19: Always Act with Honesty and Integrity - Do the Right Thing Even if No One is Watching

Honesty and integrity are principles that are inseparable from the character of a true leader. Integrity means that you will always do the right thing no matter who is watching, and honesty means that you always tell the truth. These traits are insufficient in the construction industry, but if you want to be an effective builder you have to always be honest and have integrity in every one of your actions.

Always do the right thing.

I remember a time when I was an assistant superintendent responsible for all of the site utilities for the project. I had to supervise the installation of the underground electrical, coordinate the sewer and the duct bank in 3D, and oversee the installation. When it came time to install the sewer line, I realized the duct bank was in the way of the sewer line. Immediate options were offered that would allow us to run the sewer line through the conduits and install it out of code without the inspector knowing. The installation required a two-foot separation between the duct bank and the sewer line. Instead of cutting corners, I decided to be open and honest with the project team, and I told the lead superintendent exactly what happened. He was very disappointed, but he appreciated my honesty. The fix was to install a lift station to correct the sewer line issue, and it cost $300,000 from project contingency. It was one of the largest mistakes in my career, but I was soon released from the burden of the situation because I had been honest and open with the project team, and I had acted with integrity.

Acting with honesty and integrity is an eternal principle that raises all problems to the surface through visual systems based on an established flow and stability on-site. If someone is not honest, they will hide mistakes, blame others, and will never get to the root cause that allows us to be more effective. We will always be haunted by mistakes that show up at the wrong time causing waste and variation. More importantly, a leader who does not speak with honesty and does not act with integrity will be a cancer to the team, and those types of people always end up being demoted, punished, or terminated. You cannot allow yourself to be even a "little bit" dishonest, or to act even a "little bit" without integrity.

Honesty and integrity take practice. You will, as a master builder, make many mistakes. After realizing your mistake, immediately contact someone, bring problems to the surface, and widen in your circle in the moment. Ignore the impulse to hide your error. Do not think, even for a moment, in an interview or in an owners meeting or in a negotiation with somebody else, to be dishonest in any way. Admit to mistakes and act with integrity. Practice, practice, practice is the only way to establish the right character to be a leader on-site.

Here is a promise that I can unequivocally make for you and to you. You will never be hindered in your career, ever, if you are honest and act with integrity. You will always be able to take that next step, improve, and be able to act in a manner that will bring you success and fulfillment throughout your career. I don't know how to say it any better than that, but it is 100% true.

Period 2:
Working with Others

Step 20: Read How to Win Friends and Influence People

One of the first steps in being able to lead as a superintendent among a team, is the ability to influence people on the team and project site. These are skills that people are typically not taught in schools or in colleges. We definitely do not learn these things growing up and coming up through the trades either. There is a recommended habit for every superintendent as they begin their current role to read, *How to Win Friends and Influence People* by Dale Carnegie.

You can make more friends in two months by becoming interested in other people than you can in two years by trying to get other people interested in you." - Dale Carnegie

I remember a field director reading the book at my recommendation. After reading it, he reported to me that it has brought him more success than any other book. It improved relationships and approaches at work and even fixed a family issue that had been brewing for years.

There are certain behaviors that a superintendent must understand in order to work with people. The book outlines a commonsense approach and can be implemented by any builder in the field.

I encourage you to read, *How to Win Friends and Influence People*, and to persistently practice these behaviors in the field anytime and anywhere you encounter workers or leaders.

If you learn and implement these principles, you will immediately become more effective in your role, be happier at work, and have more influence with everyone on the project.

Step 21: Develop Resolutions about Communication

Knowing what to do on the project site is one thing, but knowing how we will carry out those assignments is another. Superintendents will be best served if they create a list of resolutions that will determine their course of action and their behaviors on a daily basis as they work through people to accomplish the mission of the team.

How you are resolved to accomplish anything determines your success in the task.

I remember a time when I was so milestone driven and schedule driven that I would let my temper get away from me. I would say things to trade partners that I wasn't proud of, and I would behave in, what I was told, were unpredictable ways. I would use language that was not acceptable to others and needless to say, my reputation was not served well by this behavior. I remember following the example of a very trusted mentor in my life which entailed listing 14 resolutions for my behavior on the project. I wrote down how I would communicate, how I would solve problems, and the daily things that I would do in order to maintain a good character, a good and healthy working approach, and proper methods to accomplish the things that I knew needed to be done. I referenced this list every morning and immediately started to see the success. Everything about me was designed to generate the greatest amount of harmony and happiness for the rest of the project team and ultimately, for myself. Below is an example of those items:

- No cussing.
- No accusing others.
- NEVER, NEVER get angry again!
- Stay moral.
- Stay outside of the box with everyone.
- Message company and leadership positively. No dissention, lack of unity, or criticism amongst people.
- Take the time to be with people.
- Give service and content first. Do not take it.
- Do not waste time worrying about failure. Fail forward.
- Bring excitement [high energy] and love to every interaction.
- Keep confidences.
- Get to know first what others need and help them feel safe.
- Smile.
- Do not stress about hard things.

In construction, we are so focused on what we need to do that sometimes we forget how we were going to do it. Writing down a list of resolutions on how you are going to actualize your rules will be key to interacting with a team. Also, this is one of the best ways to self-correct and encourage new habits of behaviors. Have you ever had feedback in a performance review that keeps coming up year after year? Have you ever had something that you know is holding you back but you haven't been able to correct it? Keeping a list of daily resolutions that you reference and repeat verbally every day will help guide your behavior in the right direction and make the changes that will prevent repeat conversations suggesting correct behavior.

Create a list of resolutions for yourself that will support your future success in your current role. Make a habit of re-emphasizing this daily. You may want to consider discussing your resolutions with colleagues and friends so they can help hold you accountable to these behaviors.

You will become the person that you want to become through these proper behaviors. You will design yourself after you reinvent yourself. You will have the amount of success based on the quality of the resolutions that you've decided upon. Your role and your leader's standard work will be carried out in a way that will win friends and influence people.

Step 22: Learn Specific Applications

The days of superintendents without computer skills and an inability to use certain applications are over. In order for a superintendent to be successful in the current market, he or she will need to be proficient with technology. The use of Excel, Bluebeam, and P6 are essential in the current market. The best way to approach these skill sets is to consistently and systematically learn them in your experience as a field engineer, assistant superintendent, and as a superintendent. They will benefit you with almost every task you have in construction management.

Superintendents should be just as proficient with technology as the project manager.

I remember a superintendent who was very proficient in leading his projects in an organized fashion and communicating visually. He had some of the most well formatted logistics plans I had ever seen. When I asked, he told me he used Excel to create these drawings. I was surprised by his answer and looked at some of the things he was doing to create the graphics on these sheets. I was interested to learn that Excel has a number of drawing features and commands that allowed him to fully coordinate his project site visually on a map. Excel is not only used for finances and calculations but can also be a very handy scheduling tool. Creating bar charts and other forms of schedules and lists is very easy and effective in Excel. Excel, as I mentioned, is also very useful when creating logistics maps and other types of visuals.

Bluebeam is really the industry standard for construction PDFs. Some of the best coordination drawings, site maps, phasing maps, and other tracking tools are done and shared within BlueBeam.

P6, or another CPM scheduling tool, is absolutely necessary for a superintendent. I would caution you not to delegate the creation and maintenance of a schedule to another professional, a scheduler, or a third-party. The best superintendents are able to create their own schedules within the software, even if they don't do it long-term because of capacity and differences in roles. I would suggest purchasing a book on CPM scheduling to learn the software applicable to you and your company.

One-on-one training with your company, LinkedIn learning courses, professional online learning courses, or project team members with experience can help you learn about these applications. The key is to time block time in your week for learning and to learn and maintain the skills needed for you to be successful with technology and use it to your advantage. Your proficiency will make certain that you won't have to wait on others, allow you to communicate better as a leader, and have the skills to move fast when you are in higher level positions.

Step 23: Design your Signage, Trailer, and Project

For someone coming onto the jobsite the layout and design of signage should be easily navigated. Visual management of signage, the layout and design of your trailer, and the layout of your project for your site logistics map is very important. The appearance, the visual indicators, the sounds, and even the odors on your project all contribute to an image that shapes the culture and guides work on-site. Do not underestimate the need to intelligently design your environment. One of the first things you will do as you approach your NTP is to initiate site signage and your trailer. Make sure your entire project is designed in a way that creates good material supply chains and effective work that can be done in a flow.

Design your environment to support the project or your environment will hinder you from running your project.

Visual management is a strategy to communicate effectively with both the workers and those visiting the jobsite. I have seen projects that are poorly designed from a signage and trailer standpoint. It is a burden to navigate the site and directives through signage that is unclear. Often, however, the trailers are under budgeted, there is very little signage, and there is little thought to where things should go on the project site. There is no support from those logistical items that enables the team to work more effectively. Conversely, I remember an eight wide trailer which had three of the trailers dedicated to a craft lunchroom and a craft bathroom. The other five trailers had an open space used by the entire project team and well-designed conference rooms. The trailer was nicely decorated and organized in a way to encourage a good flow of communication and an ability for the people to effectively do work. The signage throughout the project was easy to navigate, was clear and concise, and served the team well. Everybody knew where they were supposed to go. In addition to the cleanliness and organization of the site, there were seasonal holiday decorations and music was played at appropriate times. Everything was designed to provide for an engaging and happy work site. The owner came to the project site one day and said it "felt like going to Disneyland."

When initiating your project signage, there are two key things that need to be considered--the site signage drawing and the site logistics plan. The signage drawing should be aligned with company branding standards and coordinated with the team as well as the owner on your project. This will allow a common communication system to direct workers and anyone on-site. This is very important. The second thing you will do is to create a site logistics plan that will be dynamic and usable as a tool throughout construction. We are not looking for a site logistics plan as a token drawing. We want an actual tool with intentional design based around flow and material access. The third and most important thing will be the design of the trailers. You cannot let the trailers come out by happenstance. They need to be purposeful, sized, and laid out. Desks, meeting rooms, and the layout of the space should have a positive and integrated feel. The design of what things will go on the walls should be intentionally thought through with the rest of the team and again, not by happenstance.

I recommend three different drawings as a part of your planning and preparation. Create your site logistics map, your signage drawing for the site as well as a drawing which depicts the inside of your trailer with the rest of the team. If you don't have a plan, you will become a part of someone else's. There is little benefit in becoming a part of someone else's environment for a long duration of time when that environment does not support your work.

Every person on-site needs to have the know-how, the motivation, and support systems when it comes to visual management. If you have the right support systems with your trailer, site logistics, and your signage you can have a happy and productive workforce with support systems that encourage behaviors that will eventually determine, support, and maintain your culture.

Step 24: Work on Your Schedule Daily. Sketch Everything Out

Every master builder spends time in the schedule on a daily basis. All the best builders carry their schedule with them--whether it be on a tablet or printed or written plans carried in their pocket. Schedules are key to observing project conditions and overall project plans. If a superintendent is in his or her schedule daily, it can be used as a trigger to delegate assignments.

The schedule is a superintendents' main tool. It is as a hammer to a carpenter. The job of a superintendent cannot be carried out without it.

I remember a time when I was a new assistant superintendent working on an eighty-million-dollar construction project. As I progressed in my role I stumbled onto a little habit of being in the schedule on a daily basis. There was a software I became familiar with from my BIM days called Snagit. This software was easy to use. I could hit the "print screen" button, immediately email the screenshot (called a snippet), and include notes. I got into the habit of taking snippets of the schedule and emailing instructions to trade partners and other members of the team. Some of them were reminders to prepare meetings, others were reminders to prepare certain types of equipment on a certain day, and others were confirmations for the right amount of manpower. This became a successful system for me to trigger the right things to be done at the appropriate times.

A superintendent must use his or her schedule effectively--always remembering that it can be an indispensable communication tool on the project site. The schedule is the ultimate end-all for success because it shows everyone what winning looks like on a daily basis and re-emphasizes the need for the preparation of future work. Being in your schedule and sending out assignments in that daily 30 minutes to an hour time span is one of the sustainable habits which will determine the success of the superintendent.

Carve out time every day according to leader standard work to review the schedule. As the superintendent reviews the schedule, there are six main categories that need to be considered and are as follows: manpower, materials, information, equipment, safety planning, and making sure that all quality planning is done. If everything is looked at from this broad and thorough scope, the superintendent can make assignments uninterrupted and be effective on the jobsite. The superintendent should get a software or a method whether it be by phone, text, or email, to trigger other people to complete assignments.

I can guarantee that taking a daily opportunity to study the schedule will result in an ability to stay ahead of the work, planning and executing work properly, and never being behind the eight ball. You will always be mentally on top of your schedule which will help you with your sense of urgency and drive.

Step 25: Start Field Walks and Trigger Follow-Up

One of the most successful ways to implement this principle is to take a field walk or a reflection walk at the end of the day to observe the conditions of the project and to make decisions that will determine success. Remember--through these observations, assignments can be texted, called in or emailed to become part of the trigger system for things that need to be done on-site. The most remarkable part of this is that the worker huddles the following morning can include any needed assignments to be carried out for that day.

The building will talk to you. All you have to do is take time to develop that relationship through field walks.

As a field engineer, I used to take afternoon walks on the jobsite with a superintendent who would calmly and quietly ask me questions like "Jason, what do you think of this area? What does the building say to you?" I would ponder and make observations, and if I wasn't distracted, I could come up with the right answer. We would be able to determine if crews were working too fast, if the area was behind schedule, or if certain quality expectations were not being met. He would always come back to the office and make assignments for the next day. Everyone always knew exactly what they were supposed to be doing. His system stuck with me and I continue to practice and benefit from it with great success.

Always be ahead of things in the field. We have to physically be present where the work is being done in order to see what's really going on. We can observe workers who are actually in the building so that we can notice waste and wasted motion, but sometimes as superintendents, we are distracted by correcting safety items, being in the moment, and asking and answering questions. It is a very good habit to quietly walk through the building to discover the things that you otherwise would not notice in the chaos of the day. This develops a feedback loop. Every day there is a plan and then every day there is a follow-up walk to see how we did on that plan. Corrective actions are then communicated the next morning to continuously improve. This is a system that works everywhere we try it.

Carve out a standard time every day for you to walk your area. It is not a good habit to let it happen by chance or when it's convenient because then you will end up doing your walk at the tail-end of the day and you will not arrive home on time to your family. What you need is time to observe the building so you can take pictures and/or text out assignments for the next day. Always make sure to take pictures so that you can report needed corrections to meeting agendas and trade partners in the morning huddle. By doing this--in addition to the schedule and the reading of your drawings--you will create a system that will trigger you to delegate and make corrections throughout the entire project.

The building will talk to you. You might prefer to call it inspiration, intuition, or simply reflection, but you will know what to do at all times. Others will commend you by saying, "She (or he) always knows what needs to be done." They will want to follow your example. You will have a feedback loop that will allow you to create a remarkable project on a daily basis. Problems will be fixed faster than they are currently, and you will be able to have stability throughout your project. You'll be able to appropriately lead by following the lean principles of going to the place of work in the field at all times, and you will have the amount of intuition you need to anticipate problems and win with future upcoming work.

Step 26: Return All Emails, Texts, and Phone Calls

You cannot lead others until you can lead yourself. You cannot encourage discipline in somebody else until you can discipline yourself. You cannot provide directions to a place that somebody should go if you've never been there. You must have a habit of personal discipline. Superintendents who return emails, return texts, and return phone calls are some of the most highly respected superintendents in our industry. It shows a level of respect and discipline that is admirable and also permeates every aspect of their work.

To encourage discipline in others, you must first be disciplined yourself. To lead others, you must first lead yourself.

I remember a time when I was participating in a very important proposal for a company I worked with. I interacted with a director who was in charge of marketing and she and I worked very well throughout the proposal. At the end, and after we found out that we had not won the proposal, I promised to return a phone call and discuss the aftermath of the news with her. I neglected to call her back and that left such an adverse impression on her that our relationship was never the same. I realized at that time that I should be responsible and always make good on my promises, maintain my personal discipline, and always do what I say I'm going to do.

Having discipline will encourage your success. If you promptly reply to emails, texts, and calls you will be seen as attentive and accommodating and better able to meet your career goals. If you meet your career goals, then you will have more money to take care of your family. If you have more money to take care of your family, you have a better chance in life to be happier in the field.

Be responsible with communication. Actively respond to messages. One little trick is to switch the filter on your phone to missed calls at the end of the day and scroll down through those red numbers and return those phone calls every day. Do not click on texts that you cannot immediately respond to. Leave them marked as "unread" and then respond to each at the end of the day. If you promise to respond by email, phone, text, or any other means always follow up on a daily basis. If people know you will call them back, you will be able to build trust and rapport with them.

If you do this, you will build trust with everyone, and you will not miss out on opportunities.

Step 27: Start Going Home in the Right Mental State

Everyone should be able to go home in the proper mental state. At the end of the day on a construction project you will not only be exhausted, but you might be a little bit grumpy as a result of trade partners or workers who have pushed you to a limit where your patience has worn thin. You will likely not have as much energy as you did when you started the day. It is important to reset on your way home, decompress, resolve any residual issues in your mind, and walk through that door ready to give 100%.

Your family deserves 100% of you. Be ready before you walk through that door to give them that every day.

I remember the story of a field engineer who went through one of my boot camps. He had learned in that class to take care of both his personal health and his family so he decided to set resolutions for himself to do so. He told me that his old habit was to come home and watch TV and sit on the couch while his wife made dinner and tended to the children. After he made his resolutions he decided to take a more active role when he came home at night. He stopped sitting on the couch watching TV and started changing diapers and helping with the kids and helping to set the table. His wife and children were so surprised and so much happiness entered into their life as a result of his newfound habit. He reported this to be one of the most important changes he's ever made in his entire life.

Once you sign up a person for a job in construction, you have essentially signed up his or her entire family. We need to make sure that people, workers, managers and anyone in construction can come home and walk through their door ready to take care of their families. This is the only way to have happy and healthy human beings. I can come to work and be at 100% capacity and potential and the opposite should not be true at the end of the day. We have to get into the mindset to not decompress at home, but to decompress on the way home. You should be ready to give 100% before you open that door. If you do that your relationships will be better, your emotional, physical, and spiritual needs will be met, and you will have a better support structure for coming to work. The importance of a supportive family that has bought into what you're doing cannot be understated.

Find a song or a podcast or some form of media on the way home that will allow you to decompress. A calming or energetic song or some ritual before you leave your vehicle is recommended. A very good practice is to get a song from one of your playlists and listen to that in the last five minutes of your drive home. On the way into the house jump up and down a few times, shake your hair out, and go through that door with a determination to take care of your family and to do at least one or two things to show your family how much you love them. When you have determined to go home and take action to decompress--whatever form it may take--your family has their needs met and will have extra time and energy to take care of yours. You'll find more success at home and consequently, more success at work.

If you do this, you will garner so much good will with your family, they will not even know who you are!

Step 28: Stop Working Over 55 Hours

Superintendents can work less than 55 hours during the week and still supersede expectations as the leader on their project. Superintendents need to reduce the amount of time they spend on their projects in order to increase their effectiveness, their organization, and their ability to delegate and leave.

Working too many hours is masking waste and ineffectiveness.

I remember a time when I was a lead superintendent on a project and I made a promise that I would leave by a certain time every day. I realized that the things I typically wanted to do were always getting done by forcing myself to act within a certain time constraint for the most important things. The other, more non-essential things that were left undone were left undone again the next day and the next day. I started to realize I could remove the things left undone from my to-do list by delegating them. By forcing myself to work between 50 and 55 hours, I was able to focus on the essential and reduce any waste for my role. It also forced me to delegate and leverage the skills and ability of the team through whom I worked.

Superintendents who have to work too many hours at the expense of their family time either don't have enough coverage on-site, or they do not know yet how to perform within the role. The necessity of having to work no more than 50 or 55 hours will force the superintendent to become more effective in their personal organization and be more focused on their work. He or she will stop babysitting for others, learn how to delegate, and reduce waste. If there is no need to reduce the amount of hours that somebody works, there is no incentive or force that will drive the behavior to become more effective. Working too many hours masks waste and does not serve us well. Too many hours enables us to do a lot of mediocre things at the expense of the essential--one of them being our families.

Immediately adjust your weekly calendar and your leader standard work to ensure that you do not work more than 55 hours per week. By holding to this discipline you'll be forced to adjust, delegate, and learn the things necessary to fully execute your role working reasonable hours..

There are few promises I could make to you more impactful than this one. Limiting your hours will literally shape all of your habits in your role. Waste will need to be removed as you constrain yourself to a work schedule. It will drive you to be better, to delegate better, and to organize your work. You will learn to do the work of two or three people.

Step 29: Keep Your Head Above Your A$$

One of the biggest difficulties in developing a superintendent or assistant superintendent to progress into lead positions and eventually to the general superintendent position, is to get out of the focus of doing, and into the focus of leading. We need industry superintendents to lead, plan, and execute work. We do not need superintendents who are distracted with retrieving materials, doing physical work, and being too busy to stay in their leadership role.

Keep your head above your backside.

I remember working with a new assistant superintendent who came from a craft level position. He had received training and thought he was ready to take his next step in construction. Instead of choosing to succeed, he did two things which were counterproductive to his learning path. One was to stop advantageous supervisory training because he said he was, "too busy." The second one was an inability to focus as a leader on the project. He began taking trips to the local hardware store to retrieve materials, or go off-site to check on form work, or escape to the office instead of running work in the field. He would do specific tasks, start to put his bags on again, and end up not focusing on planning the work so that it could be executed effectively. After the course of a few months he voluntarily gave up the position because it was too stressful and he went back to being a skilled craftsperson. He chose to be unprepared for the challenge of leadership because he was too focused on doing things instead of leading.

We have to keep our heads up which means watching, leading, communicating, coordinating, and correcting. If we are sweeping, physically laboring, and doing things like running to the store to get materials, we are running away from our work. We have to be in our stations and leading other people at all times. Leadership is about delegation, communication, and doing the hard things of coaching and correcting people. If we have people distracted from this role, they are not acting as superintendents. Another thing that can happen to a superintendent is that he or she becomes too busy answering RFIs, doing the trade partner's coordination, and/or running around solving problems that other people could solve themselves. We have to keep our superintendents heads up and have them delegate and assign tasks to others when appropriate so that they can watch for safety, quality, and cost. If a superintendent is not watching what he or she needs to watch, things will spiral into chaos.

Clearly identify your role as a superintendent. What are people counting on you to do? What are the things that only you can do? Once you have established your identity, identify the things that you should not be doing. Included in this list should be things like coordinating others' work, answering RFIs that the field engineer or project engineer should be answering, physical labor, material runs for anyone on-site, and paperwork that takes you away from leading from the front and having a safety presence in the field. When you become a project superintendent and a senior superintendent, you will need to spend more time in the office, but your current role requires a lot of onsite presence. Only do the things that you have to do in your role, and delegate everything else. The superintendent must be watching the project.

If you become familiar and comfortable with your role your project will be operationally excellent and you will be able to effectively plan and prepare work for others to do in an efficient and productive manner. If you do not do this, trades will not be prepared nor will workers. Issues on the project will not be corrected, and you will not be successful in your role as a superintendent.

Step 30: Become an Ideal Team Player

The most important thing for you to be successful among your team is to be an ideal team player. Ideal team players are humble, hungry, and smart. Humble means that you are willing to do what is needed for the team, hungry means that you have an appetite for success which drives your passion and energy, and smart means that you are insightful and astute when it comes to people. These are critical to your success amongst the team. Without them you will not have the influence you need to be a leader.

Ideal team players are humble, hungry, and smart.

Throughout the last few decades one of the most recurring themes among superintendents is a lack of emotional intelligence when it comes to working with people. Being smart means that we are smart with people. Superintendents have to learn this in order to be successful.

Every time a superintendent has made an intentional focus towards learning the art of dealing with people he or she has become successful and is able to progress in their career.

Good companies will use the model of humble, hungry, and smart in hiring, assessment, and coaching. You can get ahead of this curve and leverage this ability--even if your company does not use this model. *The Ideal Team Player* by Patrick Lencioni is a book that can help you as a guide. There are also tools and assessments that you can use at www.tablegroup.com that will allow you to see where you are strong and how you can make progress where you need improvement.

After you read Patrick Lencioni's book, do a self-assessment and work on your weaknesses. Ask for coaching to help you become an ideal team player.

If you can be an ideal team player with your team, one who has influence with others, you will have the support and training from the team and you will go farther in your career than you otherwise could have.

Step 31: Focus on Foresight and Planning

A superintendent's job is to see the future. This has to come from foresight and planning. A superintendent has to have a firm grasp of the present and always keep an eye on the future. A focus on foresight and planning will make the difference between a builder and a broker of trades.

Superintendents see the future.

I remember a young superintendent who had just come out of a builder boot camp who began to make visuals and sketch out schedules for the project team to look into the future. He informed everyone about where the project was headed and what the next immediate needs were. He used BlueBeam for visuals, sketches, Excel, and ultimately ended up learning P6. On his present project he can be observed knowing where his team is, has communicated that to everyone on-site, and has visuals to back it up and make the information consumable.

Find a system that will allow you to look to the future. Whether a superintendent uses P6 as the main tool, or Excel, or hand sketches on graph paper, the key is to look forward. There are many scheduling systems which will be discussed in the next book called, *Elevating Construction Senior Superintendents*, but the key here is to get started. On every project there should be a master schedule, a takt plan, a site logistics plan, phasing drawings, a basis of schedule, a weekly work plan that has been derived from a make-ready schedule and look ahead schedules, and a day plan. A superintendent, at a minimum, has to update the master schedule, provide look ahead schedules, and assist the trades in collaborating on a weekly work plan and add a day plan that can be implemented on-site.

The subjects of scheduling, the first planner system, and the last planner system are too extensive to cover in this work, but you should learn them, because you cannot detach scheduling from the role of a superintendent. Learn as much as you can from your company and from outside resources. Ask for help and for additional coaching and training. Begin with the minimum standard of updating your master schedule, providing look ahead schedules to your project site, and help your trades to create a weekly work plan and a day plan. Do not become preoccupied by which technology or medium you use. The key is to begin using anything you can to get started.

If you do this, you will become, not only a builder, but a master builder. Superintendents plan and schedule work. A superintendent who does not plan and schedule work is not a superintendent. In order to fulfill your role, you have to continually anticipate the future and see the scope of possibilities. If you practice in the system that works for you, you will know what needs to be done, be able to effectively communicate it, and keep the project on track.

Step 32: Provide Clarity and Alignment

According to Forbes statistics, over 65% of people are visual learners and approximately 30% are auditory. If you wish to communicate to the masses you need to show what you mean, not simply say it. The 7-38-55 rule in the book, *Never Split the Difference* by Chris Voss demonstrates that as humans we only gather 7% of the message from the content itself while 38% is communicated through our intonation and about 55% is transmitted through nonverbal facial expressions and body language. A superintendent's job is to provide clarity and to communicate. There is absolutely no use for a plan that is only in the superintendent's head. This is a failing system that has been used throughout construction and has always proven to be unsuccessful. Schedules are only effective if they can be communicated and shared.

Plans are only effective if everyone knows what the plan is. Strategies are only implemented when everyone is heading in the same direction. A superintendent, and for that matter all leaders on the project, need to understand that their main focus is to define and provide clarity and to align the entire project team towards effective communication.

A plan is only as successful as it was communicated. Get everything out of your head!

I remember a project that utilized a takt plan. After the first initial push back by the team to use this single page scheduling system eventually all of the trades bought into it. Every foreman on-site had an 11 x 17 laminated copy of this overall project schedule in the format of a takt plan. They were adhered to gang boxes, taped inside the hoist, and were present on every floor on huddle boards. The project always knew the end date and where each contractor should be at any given time based on that visual. If you asked what the end date was in a huddle or randomly while walking through the project the answer was December 6, 2017. The team saw as a group, knew as a group, and acted as a group.

Increase the level of communication throughout the project site through visuals, meetings, emails, and huddles. Ask the important questions: "Do all of the contractors understand the master schedule?" If the answer is no, then we have work to do. "Do all of the workers know what the plan is for the day?" If the answer is no, then we have work to do. "Do all trade partners and foremen know what the plan is for next week and have they committed to it 100% and planned accordingly?" If the answer is no, we have work to do. "Do all workers on a daily basis know what is happening that day on the project, and how they can be safe, and interact with the environment?" If the answer is no, then we have work to do. Every part of the plan, in its entirety, should be communicated to the appropriate people when it is needed. We have to use all methods of communication--audibly, in written form, physically through mockups, and most importantly through visual communication. It is the superintendent's responsibility to make this happen, communicate clarity around what the plan is, and not just to communicate it, but to communicate for understanding.

Double the amount of communication that you currently disseminate on the project site. Ask yourself what the best way to communicate is--visually, audibly, in written form, or in person? Once you investigate and make that discovery, make sure that every plan created on the project also has associated communication systems. Begin experimenting with maps, drawings on whiteboards, huddles, emails, texts, and any other available technology current in our culture.

It is powerful to have an entire project team informed. If six to twelve people on your project management team know the plan, consider how powerful it would be if everyone on your project site knew the plan, could follow it, and help you achieve the goal to see, know, and act as one unit. You will accomplish your goals, your standards, and your milestones with more people working towards them. And that vision will pull people in the direction of success.

Step 33: Harness the Energy of the Team

The energy of your team is paramount to the success of your project. Everyone on-site needs to be headed in the right direction at a speed that reflects the high energy of the team. That energy originates with the superintendent. There has to be a sense of urgency, an acceptable pace, and a sense of optimism and morale that will keep people working at full efficiency towards project goals and milestones.

Be mindful of the energy of your team. It literally determines the success of everything else.

I remember walking through a project that was a joint venture with another general contractor. The other general contractor had a fairly easy path to a substantial completion target, and their workers were not functioning with a sense of urgency towards a strenuous performance goal. They knew they had plenty of time and their pace and workmanship reflected as much. We, on the other hand, had a substantial completion deadline and needed to get done within a few weeks. When visiting the project site I noticed that our workers were going at the same pace as everyone else on-site, and the lack of energy was going to prevent us from finishing on time. We found a strategy that gave our workers a sense of identity, instilled a sense of urgency, and created a rallying cry that galvanized the efforts of everyone on the team.

The energy of the team must be watched and managed. If a superintendent finds that his or her project site is suffering from low energy, something must be immediately done to fix the situation. Low energy can come in many forms--low morale, fatigue, no sense of urgency, or unfavorable conditions on the project site. But we can control the energy we bring, the urgency we communicate, and the environment that we create that results in good morale. Energy is managed and should not be considered some soft skill technique that can be easily ignored.

Superintendents must create and maintain the energy of the project every time he or she is in a meeting, at their perch, or observing the project in quiet moments. A focus on energy will drive production.

Step 34: Stay out of the Box

Superintendents have to find a way to "stay out of the box." Staying out of the box is a term coined from the book, *Leadership and Self-deception* by the Arbinger Institute. Being in the box means having a mental perspective that deceives us in our leadership abilities. There are many times that a superintendent will metaphorically be backed up against the wall, and the knee-jerk response is to yell, scream, or say offensive things. As leaders on the projects, we must have better options to get out of difficult situations.

Always expect that others are doing their best. Presume positive intentions.

I remember a time on a particular project, the mechanical contractor was behind in commissioning the air handlers. We needed a preliminary commissioning and enhanced start-up of the units to distribute cooled air throughout the building. The leadership team for that contractor was not doing a good job of completing committed short-interval tasks. I took that group through the project, grew very frustrated along the way, and chose to cuss out and berate individuals on the walk. Some of the leaders focused only on my manner of speaking rather than discussing the issues and admitting their mistakes. I told them if they didn't get the air handlers up, "I was going to kill somebody." The leader of their company called my direct supervisor and I got into a world of trouble. I realized that I had lost my cool by getting "into the box," did not presume positive intentions, and therefore, prevented a positive outcome. I helped to escalate a bad situation, and in the end I didn't even get what I wanted.

When negativity in social interactions occur with others, there is a tendency for us to begin deceiving ourselves about their intentions and behaviors. This is all outlined in the book, *Leadership and Self-deception*, but briefly the problem is:

- You exaggerate other people's faults.
- You exaggerate your own virtue or rightness.
- You overstate the importance of factors that justify your self-betrayal.
- You blame others for your feelings.
- Over time, certain behaviors and justifications can become habitual for you.

The way to, "get out of the box" of self-deception is to do the right thing. Choices must be made before a reaction escalates beyond control. Choose to stop letting others control your behavior and assume they have good intentions and are doing the best they can. If you can do that, you can calm down, control your behavior, and think of realistic solutions to the problem. Staying "out of the box" will allow you to successfully navigate conversations on the project site.

Period 3:
Working in the Field

Step 35: Learn the 8 wastes and 5S

Earlier, I mentioned the 8 wastes of construction that a supervisor must know. A good way to memorize them is to use the acronym, DOWNTIME. They are as follows:

The 8 Wastes:

- **Defects**

  Waste caused by rework, scrap, incorrect and/or insufficient information

- **Overproduction**

  Waste caused by making more than is required or more than required right now

- **Waiting**

  Waste caused by wasted time waiting for the next process step to occur

- **Non-utilized Talent**

  Waste caused by failure to tap into the knowledge and expertise available in the organization.

- **Transportation**

  Waste caused by unnecessary movement of products and materials.

- **Inventory**

  Waste caused by products or materials sitting on the project site not being used or installed.

- **Motion**

  Waste caused by excess movement by people such as walking around and being on treasure hunts.

- **Extra Processing**

  Waste caused by working something over more than once or having waste in the value stream.

Once your eyes are opened to waste, it becomes an annoyance that is difficult to ignore. You, and everyone else, will be motivated to take incremental steps to remove the 8 wastes along with extenuating problems. You can then begin a lean idea system, or a continuous improvement system to change the project for the better. Another system that you can use is the previously mentioned 5S, the lean system of organization that sets up work environments for success so that you can see what you need to see, know what you need to know, and do what you need to do to eliminate waste and create stable environments.

I was on a project where 5S and continuous improvement systems were implemented. I observed that note cards were given to every worker on the project to carry that included the 8 wastes on one side and the 5S steps on the other which were topics of discussion in the morning huddles. Afterwards, every worker and foreman was encouraged and contractually required to spend 20 to 25 minutes 5S-ing their areas for the purpose of preparing their day and eliminating the 8 wastes in construction. Crews would remove unneeded materials, sweep, wipe down, and shine the cleaned areas, straighten and sort tools, and standardize systems to make it easier on the workers and the foreman during a 20-25 minute daily crew preparation huddle where PTPs were filled out, jobsite areas were walked, and work was made ready for the day.

Good habits begin with 5S-ing and the identification or removal of the 8 wastes. Every worker on-site should be able to identify the 8 wastes as they're working, to let the waste they see anger or annoy them, and to be able to make incremental improvements on the project site daily to remove those 8 wastes from their work. When this is done productivity will increase, throughput rates will quicken, and morale will improve. This can be reinforced in morning huddles, at orientations, and in standard meetings. Once the foreman of the workers become familiar with the system, it really becomes a remarkably self-sustaining environment of cleanliness, organization, and safety which enables all other lean efforts on the project site.

Create and print cards for 5S and the 8 wastes and distribute them to all workers for discussion in the morning huddle. Pay for worker huddles in the work orders and the 25 minutes for all crews to prepare their day and 5s their areas. Create a continuous Improvement system using Paul Akers' "2 Second Lean" method and collect videos and reward people for showcasing these videos on a monthly basis. If you can find a way to incentivize the learning and implementation of 5S and eliminate or remove the 8 wastes on your project site, you will be able to build a culture where continuous Improvement is a part of doing the work instead of a separate process. For more information on this system, read 2 *Second Lean*, by Paul Akers.

Step 36: Develop your Battle Perch

Every superintendent has to have a battle perch on the project site. This is a place where he or she can overlook the project and notice the energy of the team, the flow of work, and trends with the project site. All great generals in history had something like this from an elevated position to observe the battleground, and we need that same thing today in construction.

Every superintendent should have a place to overlook the project and reflect.

I remember a superintendent who used to climb the tower crane on his project in order to get a sense of the health and stability of the project. Each day he would climb up to the top of the tower crane and just watch for about a half hour to an hour every day. He would watch for energy, flow, and safety issues. While up there he sent out assignments and observed things that needed to be done on the project site from a high-level.

Every superintendent should have an elevated place for reflection--whether on top of a trailer, on an existing building, or a top of a crane. It is well worth the money and time to design and set up a perch because being able to observe current conditions from a high level is crucial to properly leading the project.

Find your battleground perch. Design it, use it, and reflect on things that would need to be done to make your project remarkable. If you do this you will always see what you need to see. It is easier to anticipate potential problems, and you will be able to intuitively know how the project is doing and make course corrections.

Step 37: Actively Practice Feedback and Accountability

Anyone who wants to be a leader in construction needs to practice feedback and accountability. Providing feedback in the moment is not an easy thing to do. Most of the time, we rely on getting angry to trigger a vocal reaction. This is common in the construction industry. Superintendents get pushed, feel defensive, feel the need to say something, and then get angry which finally enables them to speak up and say what they need to say. But, by that time, it is too late and his or her influence has likely been lost.

The ability to provide feedback and to hold others accountable is a rare trait, but one of the single most important skills or abilities of a superintendent in construction.

I remember a team meeting where people were reticent and did not want to engage with one another in healthy confrontation or feedback. We started an exercise where each person had to give a bit of feedback to the person next to them. For a week or two this feedback was somewhat benign and not very effective. After one particularly trying week, the team was asked to go through this exercise again and people began, somewhat nervously, to provide feedback to other people on the team. It was surprising to many on the team how easy it was to give feedback and how willing others were to receive it. Once that realization was had amongst the team, they more freely began to provide feedback and to hold each other accountable. People were also more willing to ask for feedback to be accountable to their team. They just needed to see that it was possible to intentionally practice.

Superintendents get to do the things that nobody else wants to do. Superintendents manage people, conflict, and have to be open and honest 100% of the time without being mean or disrespectful. This takes courage and can come at a high price for the superintendent. He or she must intentionally practice and develop the skill of being able to openly and honestly provide calm and respectful feedback at any given moment. Also, it is a superintendent's job to hold others accountable in a positive and productive way. There can be no doubt that the superintendent must learn this skill immediately in order to be successful.

Read the best books and take the best training regarding crucial conversations, radical candor, and accountability. Then intentionally practice what you learn in meetings. It will take quite some time to find your own style, but the point is to practice, and after years of practice it will begin to become a habit.

If you do this you will be an indispensable leader. You will be counted on to do the hard things, and you will be promoted to higher levels of leadership because you are willing to do the things that leaders at high levels must do.

Step 38: Begin Delegating

One of the key habits a superintendent needs to develop while in that role is the ability to delegate. Superintendents have historically had a difficult time delegating work so that they can become more effective. The habit and tendency has been to continue doing all the work themselves and to work too many hours. The switch from this to a more sustainable system of distributing and delegating tasks is what it will take to move into more senior roles as a superintendent.

The three most important things for assistant superintendents and superintendents to learn are delegation, urgency, and visual communication.

I remember talking to an assistant superintendent in charge of concrete and inquiring how his family was doing. He looked at me sheepishly and explained that he had not been as diligent in taking care of his family as he should. His excuse was somewhat typical as he explained that he had little time to do anything besides work. My first question to him was to ask what work he could delegate. After five minutes of contemplation, we were able to identify a minimum of ten key things that could be successfully delegated to free up time in his work and personal schedule. He simply had not taken the time to think about the things that could be delegated.

We should learn to manage our schedules in such a way that we can place everything we do into four categories: Do now, plan, delegate, and stop doing. This is called the Eisenhower method. I would recommend this method to any superintendent who is developing their leader standard work and seeking to increase their effectiveness. The key is to practice delegating, and the intentional way to do this is for a superintendent to stop doing anything he or she doesn't have to do or should not do.

A personal organization system can be of great benefit when using four categories to do, plan, delegate, and stop doing. At least one-third of an assistant superintendent's tasks and half of a superintendent's tasks can be delegated or canceled. If the paradigm switches from "what can I do?" to "what can I only do?" then the superintendent will have greater effectiveness by successfully delegating.

I have never seen a superintendent who was successful who also did not delegate. Delegation brings more hours at home, more planning on the jobsite, more personal and in-depth conversations, and raises the level of effectiveness at all levels.

Step 39: Start Sharing Everything Transparently

Superintendents are not transparent enough. Superintendents must switch into a mode of responsible transparency which means copying others on emails, sharing information, communicating intentionally, and always telling customers and your own team about actual on-site situations. Problems should not be kept secret, issues should be shared, and plans should not remain only in the superintendent's head. A superintendent will benefit if he or she is an open book in all cases.

Transparency is a strategic advantage. It is the difference between an army of one and an army of many."

I remember working with a superintendent who would not share things transparently. He would always keep problems to himself, not reach out for help, and when problems came up, he dealt with them alone. Most of the time these problems became larger issues which impacted the overall project schedule. It was painful and difficult to get information from him. Eventually the project team began to look at him and his performance as a liability to the overall project success.

Transparency is a communication tool. It is a way of thinking, a way of living, and there can be little trust without it. Non-transparent environments are environments without trust. They are environments where people deal with issues individually without the wisdom of a team. One of the 8 wastes of construction is not leveraging the wisdom or experience of the team. Non-transparent environments and non-transparent individuals perpetuate this waste and are responsible for most of the deviations in construction.

Begin copying people on emails to share information openly, to share calendars, work goals and family needs. When issues arise, practice telling the team immediately and asking them for their help. When situations arise and you form a plan, make your next step one of immediate communication to someone who can help you with your plan. Have documents and emails reviewed before they go out and make this a sustainable habit that is intentionally practiced on a continual basis.

If you do this you will be one of the most trusted partners on the team. Your career will flourish in ways that you could not possibly imagine. You will not have to shoulder burdens and problems and issues and mistakes alone but will do so with the strength of the team. You will have a better work/life balance, your stress will reduce, and you will have better insight to solving problems as you work through your career.

Step 40: Start Holding Effective Meetings

Every superintendent should be able to lead great meetings on the project site. Effective meetings are led by intentionally creating drama, thereby making it interesting to attend. The key for me is to make sure that I plan ahead of time, keep everyone's attention, and keep discussions pertinent to the people attending. A superintendent needs to focus on leading remarkable meetings for every level of the organization if he or she wants to be effective in the field. Reference the book, *Death by Meeting: A Leadership Fable...About Solving the Most Painful Problem in Business* by Patrick Lencioni to learn how.

Leading great meetings is the way you communicate; it is not a way.

I remember a time when the trade partners of the research laboratory were giving us a hard time about the number of meetings we had. There was a carpenter who was in charge of doors and hardware on the project site who was really mean and negative about operational control. We had talks with his supervisor and eventually, he started attending the meetings. We promised him that if he would engage and explain what he needed in those meetings he would find value in them. In the beginning, he sat in the corner and did not participate much, but one day he spoke of something that he needed. It was immediately taken care of and increased the flow of his work in the field. After that he became one of the most trusted trade partners on the site--in fact, he would do more than other contractors. At the end of that project, he said he didn't know if it was the best thing to have as many meetings as we did, but he very much agreed that they added value and helped to maintain operational control.

We have to remarkably run certain key meetings as field superintendents. We have to make sure that we are prepared, keep people's attention and focus, and hold discussions pertinent to the group. We have to authoritatively command the meeting and be a leader who can be respected, garner collaboration, and win the trust of everybody attending.

The Strategic Planning and Procurement Meeting:

The purpose of the planning and procurement meeting is to enact long-term strategic planning where supervisors will update their master schedule, prepare make-ready schedules from pull plans, and ensure that they have the procurement tracking to the right dates. This will be done with the project manager and the scheduler, if applicable. This meeting is also used to combine all items from all projects into the master schedule to make sure the CPM schedule or master plan is still relevant to the status and changes on the project. The agenda for this is as follows:

- Input delays and impacts - Analyze if there are any time impact analysis that need to be formatted and distributed to the owner.
- Update last week's activities - Update the status of all activities up to the current date. Updates must be accurate and precise.
- Check for missing logic - An effective schedule is effective when all needed sequences are detailed and properly logic tied. Make sure that the schedule is actually representing the amount of work it will take to finish to the end and that float paths are calculated accurately. If float paths are not calculated accurately, then the team will not be triggered to submit a Time Impact Analysis.
- Update procurement - Updating any procurement system is critical. Procurement must be a scientific and artful approach. If we can get the materials on the project site, we can build it. The project manager

and superintendent need to be on the same page with all material deliveries and all procurement. This means there might be a need to do the following: flip through the drawings, look at the model, call certain contractors, value-stream map the supply chain, do a field walk as well as anything else that would allow you and the project manager to identify that procurement is coming at the right time.

- Analyze critical path - If you can analyze the critical path, you can know exactly what things have to be done during the month without delay. Take an opportunity to filter out the critical path and set priorities for the project.
- Check milestone alignment - All of your pull plans, make-ready schedules, six-week and three-week look ahead schedules, weekly work plans, and day plans need to vertically align with all milestones on the project from the master plan and schedule.
- Three to twelve-week look ahead review - The review of the next three to twelve weeks must ensure that everything is being made ready for information, equipment, manpower, materials, and safety preparation.

Reminders for weekly schedule updates are listed as follows:

1. Update status of JHA preparation.
2. Make sure all conditions of satisfaction are very detailed and tracked in the master schedule.
3. Update past week's activities.
4. Detail new week six-weeks in the schedule from pull plans.
5. Add in owner activities.
6. Add in pre-construction meetings.
7. Add in activity code descriptors.

8. Review material delivery dates.
9. Review LOE activities.
10. Ensure schedule log is correct.
11. Export to pdf.
12. Save to Bluebeam.
13. Check float path.
14. Format any TIAs.
15. Analyze make-ready schedules.
16. Plan for 90-day, 120-day, and long-term phase planning.
17. Export to PM software to tie to procurement log system.
18. Export XER, MPP, or native file CPM.
19. Review and/or enter in planning buffers.
20. Grade contractors for schedule update.
21. Provide monthly narrative.
22. Provide monthly summary update.
23. Review variance reports.
24. Review the critical path.
25. Review retained logic.
26. Analyze float.
27. Review risk mitigation plans.
28. Insert any new planning from either changes or new design packages or phases.

Trade Partner Weekly Tactical

The purpose of the trade partner weekly tactical is to create the weekly work plan for the next week and to lead the project on a weekly tactical basis with all other trade partners on-site. You will do this with the trade partner superintendents and foremen. The agenda for this meeting includes

- Positive shout outs - Begin with two to four positive comments to begin the meeting in an awesome way.
- Lightning round - The lightning round is described in Patrick Lencioni's book, *Death by Meeting: A Leadership Fable...About Solving the Most Painful Problem in Business,* and consists of a collection of bullet point agenda items obtained by going around the room and giving each person 30 seconds to make agenda additions for the meeting. The leader of the meeting then makes sure the most critical items are done within the allotted amount of time.
- General safety - Discuss safety as your number one value on the project site and within the team.
- Review takt plan - Review the overall takt plan to understand the progression of the project.
- Review last week - Review last week's progress and commitments and report on the percent plan complete score and variances.
- Review current progress - Review current progress to ensure everything is on track and make any needed corrections.
- Create a weekly work plan - Create a weekly work plan with the trade partners that is 100% committed to and that can be executed on a daily basis in the field with all of the trade partners and last planners.
- Review key milestones - Review the key milestones of the project that will get people thinking about roadblocks. Review the key milestones of the project and make sure that there's alignment in your meeting.
- Review roadblocks - Review roadblocks on a visual map. Systematically remove them and make assignments.
- Review Plus/Delta - Reflect with the team on the good parts of the meeting and what should be done

differently in the future to make for a more remarkable meeting.

Afternoon Foremen Huddle

The purpose of an afternoon foremen huddle is to plan the next day with all of the last planners who are on-site. Make sure that the path to success for the next day is 100% clear and represented visually for the entire project. The agenda is as follows:

- Positive shout out.
- Lightning round.
- Review roadblocks.
- How did we do yesterday and today? Check status of weekly work plan.
- Discuss what winning looks like for today or tomorrow. Review the weekly work plan. Detail every portion of tomorrow's general plan for the foreman and superintendents. This should be communicated visually and represent an understanding of general items.
- What is the safety focus for tomorrow?
- Logistics. What is coming? When? Where will it go? - It is key not to write anything on the board or allow any deliveries unless you know exactly where it will be staged.
- Inspections?
- Safety observations. Share what was found and whether the previous day's items are closed out? If not, crews should fix these BEFORE going to work. Drive them to closure.
- Ask if anyone needs permits and then point them to the person / location to get them immediately after the huddle

    - Soil disturbance/ Deep excavation
    - Hot work
    - Crane
- Address weather for tomorrow.
- Review Plus/Delta.
- Leave with the information in one location.

Morning Worker Huddles

The purpose of the morning worker huddle is to create a social group with all workers on-site, win them over to project standards, and communicate the plan for the day--most especially the safety plan for the day. The agenda for this is:

- Give shout outs.
- Request feedback from workers.
- Review plan for the day in an abbreviated format.
- Define safety focus for the day.
- Review safety observations from the previous day's reflection walk.
    - Share what was found and ask if the previous day's items are closed out? If not, crews should fix these BEFORE going to work. Drive them to closure.
- Address any owner items.
- Reiterate need for cleanliness and organization.
- Ask if anyone needs permits and then point them to the person / location to get them immediately after the huddle.
    - Soil disturbance/ Deep excavation
    - Hot work
- Crane
- Review deliveries and strategy.

- Review training for the day.
- Address weather for the day.
- Encourage crew preparation huddles.
- Reinforce the need for PTPs.

Read *Death by Meeting: A Leadership Fable...About Solving the Most Painful Problem in Business* by Patrick Lencioni. Become familiar with your standard meeting systems. Get training and help for how to run the meetings to create a meeting system on your project site which will allow you to fully implement the first and last planner systems.

If you do this you will be able to remarkably run the project, plan the work, and communicate the plan to everyone on-site.

Step 41: Begin Short-interval Scheduling

Originally, the outline of this book was going to include an entire description of the first and last planner systems, but most superintendents do not need detailed descriptions of these systems. Rather, they need to begin short-interval scheduling to milestones. It may seem out of line with current lean practice and out of popularity with most, but the best practice for assistant superintendents and superintendents is to find a way to learn to schedule and communicate their schedule to the next milestone. Then, use the last planner system, engage in pull planning, make-ready scheduling, look ahead schedules, weekly work plan creation, and the full implementation of day plans. But to begin, the short-interval schedule has to be the first priority. We have seen superintendents and schedulers effectively use Excel in a Gantt chart format to sketch out their plan. Some choose various applications. Some do their planning on three or six-week planning boards on the walls. Whatever the chosen method, it is important for the superintendent to get started.

For a superintendent, the key to scheduling is to get started. The scariest step in the process of learning scheduling is to dive in and make that first schedule. Get through it, get over it, and finish it. Everything after that gets remarkable.

I remember hosting a builder boot camp (or a superintendent boot camp) for a number of people. The first question that we asked the team is whether or not they knew scheduling practices. Everyone in the room raised their hand in affirmation. Throughout the course we learned very quickly that they did not know scheduling theory and best practice nor did they know how to use the tools. This is pretty common with all superintendents. As they began learning scheduling systems, there were two to three weeks in the process where none of them even engaged with the software. We discovered their apprehension to even make a start; however, with some very forceful motivation they did so. Everything after that was rapid learning through a succession of exercises. Superintendents need to become familiar with short-interval scheduling in a manner that is compliant with company standards, but most importantly, is easy for the superintendent to understand.

First, learn scheduling as quickly as possible. Second, do it in a way that is most palatable to the superintendent in the learning cycle. We cannot get sidetracked by process, application, or format. The key is to visualize the schedule and make progress quickly so that large amounts of information can be consumed, processed, written down on a schedule, and completed. A supervisor should not only learn scheduling but should become addicted to the process and habit with best practices. Eventually, a superintendent must learn lean practices, the first and last planner system, and the ins-and-outs of CPM scheduling. For now, the key is to get started on scheduling.

Decide on a portion of the project to begin scheduling--preferably within your scope of responsibility. Start breaking that down into actionable steps listing the description of the activity, the duration, the start date, and the sequence with the other activities. You can do this on a whiteboard, on paper, on planning boards, in Excel, or in an application of your choice. Get through that first sequence and begin to visualize your personal approach and method to scheduling. The key to scheduling is listed in these steps:

1. Go through the drawings page by page, list out and hand write the activities that need to be completed to execute to work on that page.
2. Write down the duration behind each line item.
3. Complete your research through all the pages of the drawings that you are referencing.
4. Write down those sequences with their durations and descriptions on whatever medium you have chosen.
5. Once you have completed this activity, the next step is to put everything in sequence. All key sequences should be separated by a work breakdown structure. The work breakdown structure will separate different phases or scopes of the project.
6. Practice putting that information in from start to finish in a complete sequence on whatever medium you have chosen.
7. Finally, practice ways to communicate your schedule with trade partners in order to coordinate with them. You will experiment and redefine how you visualize scheduling.

Learn to schedule and create your own success. You will be able to get ahead of any work on-site and communicate the plan to everyone. There are no successful superintendents who do not know how to schedule.

Step 42: Create Your Logistics Map

Your logistics drawing cannot just be a map you create, post, and use one time. It is a strategic planning tool that will allow you to properly control your project. It should be intelligently designed, easy to follow, nice to look at, and kept up-to-date. Every good superintendent will utilize a system to control logistics in an intentional and remarkable way.

The key to production is not the area of work, but effective material supplies to those areas.

I remember that we began using a logistics plan for day planning at a pharmacy project where I worked. On the day plan maps, we would identify where materials were to be staged, where work was taking place, and then sent that drawing to every worker in the field. Our forklift operator, crane operator, and foremen began using that map to control the site. It was so remarkable to take a stagnant plan and turn it into a daily planning tool.

Plan the appropriate items and visually communicate to control logistics with the other leaders on the project site. You will want to plan for the following things at a minimum:

- Area control
- Clear access ways
- Trailer position and organization
- General site configuration and design
- Flow of deliveries
- Traffic patterns
- Pedestrian patterns
- Site security
- New utility construction
- Equipment mobilization and demobilization

- Loading platforms
- Pumping
- Slick lines and placing booms
- The hoist
- The crane
- Equipment staging
- Staging locations
- Site signage

Get help when creating your daily logistics plan. If you do this, you will begin the planning of your terrain, which has been a habit throughout centuries with the best generals in war.

Determine the following:

- Flow of construction
- Entrances to the Jobsite (You really need 2, if possible)
- Trailer location (Always get this off-site when possible or out to the construction area)
- Trade partner offices
- Dumpster location
- Temp restrooms (Keep them grouped together NOT all spread out)
- Material laydown
- Fire lane
- Temp water and temp power
- Parking lots
- Temp fence
- Site organization
- Site signage
- Emergency gathering points
- Disaster planning
- Emergency personnel access and planning

Step 43: Safety Presence in the Field

Assistant superintendents and superintendents must focus on having a safety presence in the field. When superintendents become project and senior superintendents, more of their time will be spent in the office and at headquarters, but for now, the key is to lead from the front, be in the field, and supervise with a safety presence in the field.

There must always be a safety presence in the field.

I first learned about the concept of safety in the field because it was a focus for the company where I worked. We, as a team, took this seriously and would always ask each other the following question: "Is there someone covering safety presence in the field?" We would ask this when going to lunch, heading off to meetings, or when we saw each other in the office. It was always a focus and there was always someone in the field to keep people safe.

Workers, trades, and foremen need partnership from the project management team in the field. There needs to be supervision and a second set of eyes when work is going on. It sets the tone if leaders are out there in the field with workers.

If you do this, you will always be confident that work is proceeding safely in the field and be confident that workers will go home safely.

Step 44: Own the QC Inspections!

A superintendent owns quality control! This is something that is not delegated. Field engineers and craft foremen can help, but ultimately quality is the role of a superintendent. There is a difference between superintendents who delegate quality control and the ones who will go out and finalize an inspection walk with trades or with field engineers to ensure that work is proceeding in a proper manner. Assistant superintendents and superintendents need to ensure that they not only have a safety presence in the field, but they also have a quality presence in the field.

Superintendent's do not delegate safety or quality.

Every project I visit where self-perform is active on a project, there is a distinguishing approach between successful and unsuccessful projects. Successful projects are the ones where superintendents own self-perform, participate in quality checks, walk with inspectors, and ensure all documentation is complete. Superintendents, on unsuccessful projects, can be heard saying, "I delegated that," or "I told them," or "Aren't they supposed to be in charge of that themselves?" There's a lack of ownership which is dictated by the attitude of the superintendent, and those projects are always rampant with quality issues.

A superintendent must be intimately engaged with a project from a quality and safety standpoint. Inspections should be completed by the superintendent every time. Paperwork should be checked by the superintendent, daily. The superintendent should use the quality process espoused by the company to run work--not just as a side thought or ancillary system.

First, a superintendent should learn the quality process used by the company as a method to manage the project. Second, an assistant superintendent or superintendent should begin working with everyone on-site performing inspections and walk with all city inspectors during inspection walks.

If you do this you will have a quality project. You will set the tone, people will know how important it is to you, and everyone will follow your attitude. It is impossible to communicate a high urgency and high sense of care for quality when it is delegated with a low level of care. People care about quality when the superintendents do.

Step 45: Line out your Foremen and Craft Support

As a superintendent you must lead everyone on the project, including the foremen who work for the trade partners. They are employees of different companies, but they work for you on the project. You must have enough influence with them to teach them, line them out, guide them, and be able to provide feedback and coaching on a regular basis.

Your foremen on-site work for you, even if they have another company name on their card. Line them out.

I remember a story about concrete foremen who were not producing on a large-scale structure. The company did not know how to correct the situation so the project superintendents scheduled a task force meeting and gathered them daily for instruction to guide them through the next milestone on the project. They began performing and eventually, a qualified superintendent came to lead them. If they had looked at them as a separate entity they would have failed, but they took them under their wing, provided direction, and they hit their date.

Expect certain things from your foremen--lead them, teach them, and line them out continuously.

Here is a brief list of foremen expectations:

- Locate in the jobsite trailers or on-site.
- Bring materials to the site, "Just in Time!"
- Enforce company safety rules from the Safety Manual
- Keep crew operations clean and organized 100% of the time. Do not wait until the end of the day.
- Do not let materials touch the ground--either new materials or trash.
- Spend time every morning teaching about the 8 wastes, 5S-ing, and lean concepts. Allow crews time

every morning to get their day setup, cleaned, and organized.

- Participate as a project team member in the weekly meeting system, and at huddles.
- Encourage your crews to come up with lean improvement ideas.
- Maintain parking for your company and ensure there is no impact to the customer.
- Post material deliveries for equipment and supplies in a visible location for the entire team to see. All deliveries will be scheduled per jobsite rules.

Every foreman supervising in the field will attend a preconstruction meeting prior to the Feature of Work and prepare for the meeting by reading all associated plans, specifications, and shop drawings. The product of this review will be a trade provided checklist of critical quality items pertaining to the installation. If the foreman or superintendent do not come prepared, the meeting will be summarily cancelled and rescheduled as soon as a commitment can be made to properly research the scope.

Contractors will participate in phase planning sessions to develop the overall schedule to a Level 4 schedule.

If you use all the foremen on-site to run the project, you will be better enabled as a leader and better able to control the project.

Step 46: Make Work Fun

Superintendents need to find a way to create remarkable environments that are fun. If this can be done, you will have people working at 100% capacity.

Work can be fun!

I remember a project where foremen huddles were made into a format where people could enjoy themselves.

People would crack jokes and sometimes, the super would play funny videos, but there would always be something to make the people enjoy coming to the meeting. The team began to include more and more practices that made that meeting fun. They became so effective that the owner began flying other contractors from other projects to the site to see how the meetings should be run on a successful project.

People work best when they enjoy coming to work. An unhealthy work environment is one where productivity is low. If you can create an environment with comradery and meetings that are engaging and fun--complete with music, decorations, games, monthly barbeques, and laughter--you can turn your project from work to an addiction.

Make a fun project site. Implement the steps below and see where you can take it.

Work can be fun when people:

- find their work fulfilling. They feel like they are actually accomplishing something.
- feel appreciated. Somebody took the time to say, "Thank You!" or "Good Job!"
- feel they can be themselves and express their personalities.
- feel they are valued for their minds, not just for doing mindless work.
- like their coworkers, including their boss. Working with nice people is a lot better than working with unfriendly people.
- know their jobs are valued by the company. Fulfillment comes from within, but people also need to know that the company and the boss value the job.
- believe in the mission of their company and their job.
- feel they are treated with respect.

- have a true sense of ownership in the company and its mission. It's always more fun to feel like an owner than an employee.
- do not have to check their values at the door. Most people would prefer to work in an environment of integrity where they are not asked to do anything contrary to their values.
- celebrate success. People like to celebrate victories and they like to be recognized when they have helped accomplish something. Even if they personally had nothing to do with it, a success for the company or department should be cause for celebration.
- are not liquidated for making a mistake. Thomas Edison once said man would never fly. People make mistakes.
- are not turned into scapegoats when something goes wrong. Sometimes this takes a leader with broad shoulders who will take some heat rather than looking for somebody to blame.
- share credit. Sometimes this takes a leader whose ego does not require the spotlight.
- draw their self-esteem from what other people think of their job. It helps when friends and family think the job and the company are really cool. Don't underestimate this one.

This list originated from the book *Do the Right Thing* by James F Parker. It has been modified for construction. If you can make work fun then you will live a remarkable life and enjoy coming to work. Meetings will be fun and you will enjoy the people with whom you work.

Step 47: Learn the First and Last Planner System

In addition to short-interval scheduling, it is also important that you begin immediately learning the first and last planner systems. The first planner system is the schedule from which budgets are made and project deadlines are established with milestones. The first planners on the project (or the first people in the planning cycle) use this system to plan the project with the Takt plan and master schedule. The last planner system is aptly named for the last planners (or foremen) in the cycle and is designed to take milestones from the first planner system, pull-plan to them, and create collaborative networks of commitments in order to drive weekly and daily production work. This system is graded by what is called Percent Plan Complete, or PPC. It will be important for you to begin learning these systems from your company, industry resources, and/or from hands-on training like the superintendent boot camps offered by Elevate Construction. Below is a brief description of each step:

First Planner System Tools:

- **Establish Your Takt Plan:** The Takt plan is the overall plan for the project at its highest level and outside of your CPM software. It shows the network of phases and the flow with which you will build the project. It is important to establish your Takt plan first so you know the overall plan and focus on when you should do the work, rather than how soon you can do it. This will help identify the needed project duration and easily collaborate with others to ensure you have a general plan as early as schematic design.
- **Establish your Master Schedule:** Your master schedule is the schedule that controls all other schedules and databases. This is usually done in a CPM software, using the critical path method. This schedule is your contractual schedule and the one that you use to track progress with owners and trade partners in the

legal sense. It will be important for you to detail the schedule to the proper level with your project team and scheduler. Using as many or as few options that your CPM software is capable e.g. Enterprise Management, Resources, Cost Loading and S-Curves, EVM, SPI)

- **Establish Your Visual Sequence Maps:** Every project schedule must have sequence maps. Without sequence maps, there is no schedule--even if you have the CPM schedule to go with it. You can only understand the master schedule with the right sequence maps. You should identify sequence maps for foundations, superstructure, interiors, finishes, exterior, and commissioning. It is best to have these built into the design breakouts to match the plan and design.
- **Begin Phase Planning:** Phase planning means that you are planning general phases of the project up and to certain milestones. This can be done in many forms, including pull planning. It will be important for you to learn how to plan phases of the project so that you can create make-ready schedules and have trade partners deliver information, equipment, manpower, and materials to the project at the appropriate date. The closer you are to the activity in question, the more detailed and accurate the planning should be.

Last Planner Tools:

- **Utilize Pull Planning:** Pull planning is a collaborative system in which you--along with trade partners--plan from the milestone backwards and make mutual commitments in a sequence of work. This form of planning allows everyone to see, know, and act as a group.

- **Produce Look Ahead Schedules:** Look ahead schedules can be very effective in planning work and in communicating to trades upcoming work in the next short interval.
- **Begin Weekly Work Planning:** Weekly work planning is part of the last planner system and is a collaborative process where trade partners collectively make commitments for the plan that will be executed the next week.
- **Begin Day Planning:** Day planning is a visual exercise used to plan out the next day with all foreman on-site according to the weekly work plan.
- **Know your Production Rates and Track Daily:** Every plan should also be filled with the right production information so crews know how much of a certain product they need to produce that day depending on crew size to achieve the weekly work plan.
- **Identify Winning Daily:** Every contractor and worker on-site should know what winning looks like daily. The crew should have the plan communicated to them on a daily basis and be told how they did the next day so that they can make adjustments to improve.

This short summary of activities for the first and last planner systems is only a brief preview of the systems themselves. This information is covered in detail in boot camps and in other industry resource training. Search out the training that is right for you and begin now to integrate it into your own personal workflow.

If you do this, you will be a master builder, adept at lean principles, and be able to run and schedule the most remarkable projects in the world.

Step 48: Press the Attack with Cleanliness

Once you think you've gone far enough with safety, take it another step further. Safety and cleanliness thrive when you strive for perfection in both systems. You must press the attack with cleanliness and not turn back. You are looking for floors and staging areas that are as clean as a manufacturing facility. It can be done, and it must be done if you want to have the ability to operationally control anything else in your projects.

Cleanliness, organization, and the right-sizing of inventory buffers are the best indicators of a project's health and stability.

The key for implementing cleanliness is to raise the mental set point of the team. I remember a project where supervisors said they wanted to have a clean and organized site. I would do walks on Wednesdays, take observations through the project communication system, and ask for certain things to be cleaned up. I would be very proactive in correcting issues with our self-perform crews and I expected perfection. After about two months, the project team became addicted to the system. I had always wondered if you could scale cleanliness expectations to other teams and I found out the answer is, "yes." They began using the same communication system to hold everybody accountable on the project site, and after three weeks, I was no longer needed to sustain the system. I have been gone for short and long durations from the project site, and if you return there today, you'll find one of the most remarkably clean projects in the state.

What should we remember on every project?

Every task, project, and effort starts with being clean and organized. This is always a constant. Taking care of this for others is masking waste. We need to expect and enforce cleanliness and organization on every project.

Why do we do it?

- Safety for trips, falls, and falling objects.
- Customer satisfaction.
- The way we care for cleanliness and organization sets a standard for all other work on-site.
- It is a key indicator for crew disorganization. If a crew is messy, they have dysfunctional leadership and routines.
- You see quality problems.
- You can see safety problems.
- Cleanliness forces workers to improve their work habits.
- It indicates whether the GC has control of the site.
- It indicates morale.
- Unclean jobs are typically jobs that are in trouble.
- An unclean job typically does not finish on time or finish well.
- The level of cleanliness is an indicator of the level of care paid to costs, quality, safety, production and all other systems.
- A clean site is a stable site. Stable sites and systems are the only environments where improvements and human respect can flourish.

How do we do it?

- Decide there will never be a composite cleanup crew or paid laborers on-site.
- Get team buy-in about enforcing a clean site.
- Shut down crews if their cleanliness is not perfect. Why do I say perfect? Because a little mess snowballs into a big one and then habits cannot be controlled. If perfect is the standard, then the site will be clean. Perfect will have to do.

- The PM, super, and team have to care deeply about cleanliness philosophically and practically. They will need to shut down crews for the first six weeks until it is a project habit.
- Seeing top leadership pick up trash and cleanup is the most powerful example on-site.
- Preach about it daily in the morning huddles.
- Take site walks, take pictures, and mention observations the next morning. Enforce, enforce, enforce.
- Everyone on-site must set the example and enforce the policy.
- The orientation should explain the approach to everyone.
- Daily safety huddles should remind people and train them on standards.

Hold the line, don't budge, be strict, calm trade partners, and in weeks the site will uphold the standard without a lot of oversight. Every new wave of contractors will have to be trained.

If you decide to press the attack with cleanliness on your site, you will never turn back. You will always run clean projects, win over inspectors, delight your customers, and be promoted for the good clean work you do on-site. How you do one thing is how you do everything, and cleanliness will literally affect every other part of your project and your career.

Step 49: Begin Using Zero Tolerance

Zero Tolerance systems on-site work everywhere we try them. The key is to not tolerate bad behaviors on-site, and to keep people safe and making money. The following brief outline will demonstrate how this is most effectively done on the project site.

The culture of any organization is shaped by the worst behavior the leader is willing to tolerate.

Every time we try Zero Tolerance it works. The key is to establish common standards, orient everyone to the standards, and decide as a team on a collective form of consequences. After that, it works only if every member of the team is committed to implementing and enforcing the rules. For the first couple of days, people will be upset and you will have to remove people from the project site. After that, everyone will get used to the system as long as you are consistent. After approximately six weeks of effort, and only then, will the trade partners begin to notice the difference in safety and advocate for the system.

What should we remember on every project?

Respect for people! That's it. That's why we do everything. We take care of the customer because we respect them, their staff, and their end users. We take care of our people because we respect them. We treat trades well because we respect them. We are safe because we respect our people and their families. We provide adequate facilities, bathrooms, lunchrooms, and treat people fairly because we respect them. We keep perfectly clean job sites because we respect the productivity of other trades. We bring materials on time, and just in time, because we respect other trades. We do not tolerate safety violations because we respect people's lives and the well-being of their families.

Why do we do it?

- Our contracts say we should.
- OSHA requires us to educate and control the safety on-site.
- Trade partners expect us to keep people safe and enable their productivity.

- We have the responsibility of making sure everyone on-site knows expectations and follows them.
- Each trade partner has their own rules for safety which we need to respect.
- It's the right thing to do.
- If it's wrong, why would we tolerate it?

If we believe everyone has equal opportunity, reasonable intelligence, and the ability to work on our sites, they can follow the rules. When we don't enforce the rules, we are effectively saying, "We don't care. You can't, won't, or are not intellectually able to follow the rules because you are not as good as me." That is not a respectful or true message.

How do we do it?

Decide on zero tolerance items:

- Any violation of safety that is contrary to the company standards, orientation, and OSHA 10 training.
- Anything that is indicative of bad behavior, bad attitudes, not paying attention, or not being trained for the task.
- Anything that is high risk like ladder use, electrical, fall protection, confined space, excavations, etc.
- If it is an honest mistake that could not have been prevented by being mentally present, having a good attitude, and typical training, I would remind them.
- Starting with safety glasses is my preference. It has a psychological effect. It sets the standard of behavior onsite. If someone will not wear their safety glasses, they will not wear their fall protection properly. The important standards will be kept like the minimum standards are. It is a mental and behavioral trigger.

Enforcement of

- On time deliveries
- Organization
- Just in time deliveries and intentional staging of materials
- Perfect cleanliness
- Not covering or leaving non-quality work

- Everyone on-site must set the example and enforce the policy.
- The orientation should explain the approach to everyone.
- Daily safety huddles should remind people and train them on standards.
- If someone is observed, you say to them, "Because I care about your safety, we need to give you time to focus, re-train, or plan the work. So, let's have you go home for the day, and you can come back tomorrow for orientation" --unless it is a major violation.
- Send an email to that person's company explaining why that person was allowed to go home for their own safety and the benefit of their family. Ask that the person is re-trained and offer for them to come back through orientation--if not a major violation.
- Log the name and violation on a log to track repeat offenders or folks who cannot come back.
- If it is minor, they come back through orientation; if they do it again, they cannot come back; if it is a serious violation that could have killed them, they cannot come back.
- Hold the line, don't budge, be strict, calm trade partners, and in weeks the site will uphold the

standard without a lot of oversight. Every new wave of contractors will have to be trained.

- If you implement Zero Tolerance on-site, you can have a remarkably well-run project. You will have fewer safety incidents and have less need for babysitting in the field.

Field Position Levels

Below we have listed the general role from Super level 6 to Super level.

The purpose is to entice you to higher levels of achievement and provide a guide for you to evaluate your progress. Some companies have modified versions of this field progression list, so make sure you are aligned with your specific company.

A) **Field Operations Director - SUPER 6 – Operational Excellence, Balanced Teams, Reduce Risk, Allocate Resources**

   i) Role:

      1) Ensures projects are safe, clean, organized, with high morale
      2) Develop/Direct
         (a) People
         (b) Processes
         (c) Resources
         (d) Teams
         (e) Operations implementation in field
      3) Responsible for several General and Project Superintendents, General Foreman, Laborers/Craft. Responsible for field operations, allocation of staff, work processes, interface with scheduling, pre-construction, survey, BIM, and other departments
      4) Roadblock remover for resources

5) Improves all aspects of field operations within business
6) Helps struggling projects and project teams
7) Runs Superintendent meetings
8) Meets with Directors
9) Executive level management
10) Communicates company's goals and vision
11) Supports development and implementation of operations initiatives
12) BD support in project pursuits for
    (a) Logistics
    (b) Phasing
    (c) Schedule

ii) Skills:
1) Capable of managing large, complex and multiple projects over $250 million. 20+ years' experience
2) Executive level skills
3) Advanced problem-solving skills
4) Lean champions
5) Field Engineer champions
6) Construction methodologies champions
7) Communicates well with Project Directors
8) IPD champions
9) + SUPER 5 skills

iii) Emotional Intelligence:
1) Advanced team building skills
2) Proficient with training
3) Master culture builders

iv) Outcomes:
1) Projects are safe with a presence in field

2) Projects are clean and organized
3) All departments continue to support projects more effectively
4) Teams are built that work well together
5) All Superintendents are being trained well
6) The role of General Superintendent is functioning well

B) General Superintendent – SUPER 5 – Build Teams, Oversee Projects from Start to Finish, Predictable Results [Leader]

i) Role:

1) Ensures projects are safe, clean, organized, with high morale
2) Oversees multiple projects
3) Solves problems with staff, manpower, and materials
4) Roadblock remover for projects
5) Manages allocation of Craft
6) Responsible over several Project Superintendents, General Foreman, Laborers/Craft. Responsible for operational stability of projects.
7) Performs monthly and quarterly risk audits
8) Assists in Superintendent and Field Engineer training
9) Develops Craft to enter salary positions or Lead, Foremen, General Foreman positions
10) Oversees field operations from pre-construction to closeout
11) Supports and improves self-perform work
12) Develops trade partner relations
13) Is current with best industry practices
14) High team building focus

15) Oversees process from start to finish on project – the constancy with the Project Director /Project Executive
16) Leads in emerging industry practices

ii) Skills:
1) Capable of managing large complex and multiple projects over $160 Million. 18+ years' experience
2) Understands all aspects of construction in the field
3) Proficient in proposals and with sales relationships
4) Project startup champions
5) + SUPER 4 skills

iii) Emotional Intelligence:
1) Team building
2) Knowledge of IPD
3) Approachable
4) Excellent communicator

iv) Outcomes:
1) Projects are safe with a presence in field
2) Projects are clean and organized
3) On-site superintendents are mentored and feel supported
4) Project transfer from pre-construction to construction is smooth
5) Every project has an effective plan to succeed
6) Every project finishes on time without a crash landing
7) Every project has check-ins that allow resources to prevent problems
8) Craft is developed and every project has competent site control

C) Senior Superintendent – SUPER 4 – Master Planner and Organizer, Sees the Future, Large Scale [Emerging Leader]

i) Role:

1) Ensures projects are safe, clean, organized, with high morale
2) Ensures all field systems are being used properly
3) Plans and coordinates large or complex projects
4) Roadblock remover for project
5) Manages and allocates manpower
6) Manages and allocates materials
7) Master planner and scheduler
8) Owner interface champion
9) Trains Supers and Field Engineers
10) Develops Craft into salaried positions
11) Responsible over several lower level Superintendents, General Foreman, Laborers/Craft
12) Responsible for steering the ship

ii) Skills:

1) Capable of managing complex and/or multiple projects over $100 Million. Minimum 15 Years' Experience
2) Increased complexity of projects and systems
3) Can oversee all phases of pre-construction
4) Can oversee commissioning
5) Milestone alignment champion
6) Complete understanding of all scheduling systems
7) Can interface with all project management systems and software
8) Can oversee entire project team

9) Learning and developing pre-construction and proposal support
10) Able to completely control project for operational excellence
11) + SUPER 3 skills

iii) Emotional Intelligence:

1) Able to provide clarity for multiple areas of the project
2) Manages people, goals and performance
3) Advanced understanding of Lean – has read *The Goal*
4) Team building strategies
5) Familiar with the *33 Strategies of War* and The *Art of War*

iv) Outcomes:

1) Project is safe with a presence in field
2) Complete project control with operational excellence
3) High performing team
4) Project is clean and organize
5) Project is aligned
6) No lasting issues with manpower or procurement
7) All geographical areas of project are aligned
8) Every group is working effectively
9) On-site supers are mentored and supported

D) Project Superintendent – SUPER 3 – Plan and Organize a Project [Executing Well]

i) Role:

1) Ensures project is safe, clean, organized, with high morale
2) High engagement in safety management

3) Plans and schedules work – holds milestones
4) Plans manpower and materials for project
5) Removes roadblocks
6) Oversees other Superintendents
7) Manages risk for project
8) Project setup and overall management
9) Manages materials
10) Owner interface champion
11) Responsible for Superintendents and Assistant Superintendents, General Foreman, Laborers/Craft. Responsible for organization, work methods, scheduling, cost control, conformity with drawings and specs, workmanship (QC) either directly or through a General Foreman or Foreman
12) Oversees Carpenters and other resources on project
13) Oversees Field Engineers
14) Ensures team is healthy

ii) Skills:

1) $50-100 Million dollar projects. 10 to 15 years' experience
2) Creates and manages a Schedule in P6
3) Operates all scheduling systems
4) Plans and manages complex logistics
5) Advanced communication skills verbally and in writing
6) Uses all PM computer systems and tech on-site
7) Strong with MEP
8) Procures and sets up cranes, shoring, and other complex construction means and methods
9) Oversees punch list and closeout

10) Oversees inspection systems, final inspections, and testing
11) Weekly schedule updates
12) Leads project Operations Meetings
13) Understands key Field Engineering principles
14) Understands how to use BIM
15) Negotiates with AHJs
16) Proficient with permitting and compliance
17) + SUPER 2 skills

iii) Emotional Intelligence:

1) Effective Team Leader – has read *The Advantage*
2) Solid understanding of Lean principles – has read *The Toyota Way*
3) Able to provide clarity for the project and communicate plan and strategy – has read *The Four Obsessions of an Extraordinary Executive*
4) Advanced communication and leadership skills and setup

iv) Outcomes:

1) Project is safe with a presence in field
2) Project is clean and organized
3) Project is on schedule
4) Everyone always knows the plan and their commitments to reach it
5) Milestones are always aligned
6) Team is functioning well
7) Trade Partners and self-perform are effective and there is high morale

E) Superintendent – SUPER 2 – Plan and Execute Work

i) Role:

1) Ensures project is safe, clean, organized, with high morale
2) Short-interval planning champion
3) Executes work
4) Milestone alignment with project schedule
5) Proactively engages in safety
6) Responsible over lower level Superintendents, General Foreman, Laborers/Craft. Responsible for organization, work methods, scheduling, and conformity with drawings and specs, workmanship (QC) either directly or through a General Foreman or Foreman
7) Responsible for production

ii) Skills:

1) $15-49 Million dollar projects. 5 to 10 Years' Experience can create a vision and strategize
2) Oversees short-interval planning
3) Can run meetings and phase planning meetings
4) Oversees weekly work planning
5) Oversees day work planning
6) Can run daily Foremen Huddles and Morning Worker Huddles
7) Can supervise self-perform work
8) Can mentor Field Engineer program
9) Uses procurement log as a tool for alignment with project schedule
10) Can develop phase/sequence plans and CPM schedules
11) Impact control planning, ICRA, ISLM, and MOP management

12) + SUPER 1 skills

iii) Emotional Intelligence:

1) Understands how to implement change – has read *Switch*
2) Fits the ideal team player
3) Can oversee and interface with Owner, Trades, Team, and Public
4) Basic understanding of Lean principles – has read 2 *Second Lean*
5) Interacts well with Owner

iv) Outcomes:

1) Project is safe with a presence in field
2) Project is clean and organized
3) All planning systems work onsite
4) There is always a plan for the next 6 weeks

F) Assistant Superintendent – SUPER 1 – Safety and Quality Presence in the Field, Executes Work [a learning Super

i) Role:

1) Ensures area is safe, clean, organized, with high morale
2) Supervises self-perform crews
3) Onsite logistics control – cleanliness and organization
4) Safety supervision – safety presence in field
5) Works with onsite crews for quality implementation

ii) Skills:

1) $5 -15 Million-dollar projects. 3 to 5 Years' Experience
2) Computer skills for email, Excel, and personal organization

3) Engages, resolves, and closes issues
4) Able to generate flow on a project
5) Is a quality champion
6) Can create a 3 week look ahead plan and implement on the project
7) Learning
    (a) BIM
    (b) Planning for safety
    (c) Learning about job costs

iii) Emotional Intelligence:
1) Able to work with Trade Partners
2) Integral with project team
3) Ability to mentor

iv) Outcomes:
1) Project is safe with a presence in the field
2) Project is clean and organized
3) Deliveries are scheduled and organized
4) Owner feels we are delivering quality
5) Self-perform is successful

Superintendent Commandments

There are certain principles and actions that every role must carry-out to effectively succeed. These are summarized and listed below as the Superintendent Commandments. Following these Commandments will ensure that you will be successful as a superintendent.

1. Start your day by prioritizing your task list. If you don't have a task list, create one.
2. Study the drawings for 30 minutes every day.
3. Review the schedule daily and send out assignments to prepare work.
4. Take a reflection walk daily and send out assignments.
5. Keep the schedule on your person when in the field. Update schedule notes as you walk the project.
6. Visualize the plan daily in your mind. Draw sketches.
7. Communicate critical components of the plan daily to everyone.
8. Always keep your tape measure on you.
9. Ask questions as a form of habit, and you will know everything you need to know.
10. Be transparent about everything. Copy on emails, communicate, and tell the truth.
11. Reach out for help from other builders and networks to be more effective.
12. Learn monthly from reading books, training, going to events, and visiting other projects.

13. Remove roadblocks as your top priority and keep work moving forward.
14. Ensure manpower, materials, and needed permissions are ready for work.
15. Start every day making sure the project is safe.
16. Always keep workers at a steady pace and the project 100% clean.
17. Return all emails, texts, phone calls. Show people that you respect their time.
18. Place attention on safety, housekeeping, project flow and energy during all site walks.
19. Give and receive daily positive communication with your project manager and work with them as your equal accountability partner.
20. Don't work more than 55 hours.

Superintendent Daily Tasks

Every superintendent must do certain things daily to be successful. Below is a list of items you can measure daily to ensure you are winning.

1. Speak up effectively in the morning huddle about critical items for the day.
2. Really think for a minimum of 10 minutes about safety and make the effort to know that there is a plan to keep everyone safe that day.
3. Study the drawings for 30 minutes and send out snippets to people of things to watch for, order, or prepare.
4. Update, modify, plan, add, or review the schedule for 15-30 minutes and send out snippets or reminders to people to prepare all needed items to start work (manpower, materials, equipment, information).
5. Take a focused reflection walk, observe flow, energy, cleanliness, and safety for the project, and text, email, or call-in assignments.
6. Ensure work is taking place with correct crew counts by area according to the day plan.
7. Review quality focus for all new activities that day.
8. Go home in a proper mental state to care for your family.
9. Do something to sharpen the axe and learn. A book, podcast, or training.
10. Work less than 55 hours that week.

Superintendent Secret Sauce

In a book about Netflix culture, there is a section that talks about secret sauce. Netflix uses this document, which is basically a PowerPoint slide with small sections of information or simple sentences, to on board their new employees, and constantly create and shape culture within the company. Using secret sauce for organizations, teams, and roles has historically been very effective in onboarding everyone to the key principles, beliefs, and actions that create a winning environment. Below are some Secret Sauce items for superintendents. A good Superintendent:

1. Sees the future
2. Doesn't work more than 55 hours
3. Drives with urgency
4. Creates good habits
5. Leads with passion
6. Communicates vision
7. Does the right thing even when no one is looking
8. Are accountable
9. Creates flow
10. Doesn't delegate safety, quality, and schedule
11. Takes ownership of the project
12. Are equal to the project manager
13. Welcomes feedback
14. Errs on the side of action
15. Are team builders

16. Brings and maintains high energy on the project
17. Fanatically follows-up
18. Returns phone calls, texts, and emails
19. Stays calm when decisions are hard
20. Are advocates for workers
21. Always protects finished work
22. Helps to make a remarkable experience for the whole project
23. Is willing to ask for advice and reach out and network with other supers
24. Respects workers, individuals, trade partners, and our customers
25. Focuses on foreseeing and removing roadblocks
26. Understands that motion does not equal value
27. Knows that overproduction and inventory is the mother of all waste
28. Knows that deliveries and procurement management has to be preceded by a solid plan
29. Doesn't keep a plan in his or her head--they communicate to create success
30. Knows that learning and improvement is crucial
31. Shows respect for others by being transparent
32. Applies Genchi Genbutsu (understand the situation--go and see)

Project Audit

You can use this checklist to help you implement the concepts from this book. Do you have these items implemented? If not, what is your plan?

Schedule:

- Team has a significant performance challenge to focus on.
- Team has a milestone or milestones they are aggressively working toward.
- Master schedule is used as a tool for triggering work and procurement and is updated weekly.
- Make-ready scheduling is done 120 days out and used for ordering manpower and materials.
- Weekly Work Plan is created weekly in Trade Partner Weekly Tacticals.
- Day Plans are created in the Daily Foreman Huddle.
- Communication is getting to the Workers in the Daily Worker Huddle.
- Day Plan is posted in one location for the entire project.
- Superintendent is in weekly procurement meetings.

Operations:

- Team has fanatical roadblock removal systems.
- Team Weekly Tactical is effective and held on time.
- Team Daily Huddle is held and effective.

- Trade Partners grade the GC weekly.
- Craft has:
    - Remarkable Bathrooms.
    - Remarkable Lunch Area.
- Team coverage is discussed daily and planned weekly.
- Quality process is managed in team meetings.

Safety:

- Safety training is being rolled out in an effective manner.
- Weekly safety walks are done weekly.
- Daily Worker Huddles are held daily
- Team has zero tolerance implemented
- Safety permits are issued and reviewed daily
- PTPs reviewed daily by geographical area
- Project is perfectly clean
- Project is perfectly organized
- Project has a daily system to correct safety, logistics, and cleanup items

Super Duties:

- Daily reports
- Observations
- Progress photos weekly

Project Expectations:

- Everyone knows how to **be safe** in their task
- Everyone knows **what they are installing**
- Everyone makes **improvements daily**
- Keep **bathrooms clean**

- Be **good neighbors** and take care of the customers' needs
- Nothing **hits the floor** - No materials, trash, or other items hit the floor
- **Just-In-Time** deliveries and scheduled deliveries - Create correctly sized inventory buffers for all materials and coordinate daily
- All **cords off the floor** and managed in a remarkable way
- Everything **on wheels**, Greenies, or painted pallets
- All **accessways clear** at all times
- Organized workspaces - **Everything clean and organized**. A place for everything, and everything in its place
- **Pull work** behind you - Nothing left behind, clear and sweep your areas, and leave a complete area

Reading

I thought it would be helpful to list some books that can be beneficial for a superintendent hoping to improve their management skills. As a part of this training, we would ask you to read the strongly recommended books in their entirety as soon as possible. The world will pay you based on what it thinks you are worth, and your worth at work is measured by your education for your position. If you want to make more money, strive for the next promotion, or be more proficient in your role, then you must get to work on this reading list.

I recommend training your mind to listen to these books habitually as you travel to work. After a short time, it will be possible to listen to the content at 1.5 or 2 times the recorded speed. At that rate, you can listen to a four-hour book during two or three days of commuting.

The culture we have is the result of the consequences of our actions, which are based on our beliefs. We will experience no improvement if we do not improve our beliefs. Those who are not willing to pay the price to read and learn will suffer the consequences. We can learn from the wisdom of others or a sad experience. We hope you will learn from wisdom and enjoy more success, fulfillment, and financial prosperity in your role by doing what others are not willing to do—read and learn daily.

Strongly Recommended

***2 Second Lean: How to Grow People and Build a Fun Lean Culture* by Paul A. Akers**

In *2 Second Lean,* Paul Akers teaches the process of implementing lean principles that have worked for him in his manufacturing facility. The concepts do not tie exactly to construction but can be applied to any leadership situation. He leads the reader through the process of creating a lean culture of excellence. He simplifies it so that anyone can understand and does it in an interesting and abbreviated form. We recommend everyone study this book twice to be sure to absorb the message.

***How to Win Friends and Influence People* by Dale Carnegie**

This book will help you develop your interpersonal skills. It was written initially as a lecture series about how to deal with people. It is well written, with illustrative stories in every section. Many have found this book life changing and immediately implementable, and we hope you will as well.

***Switch: How to Change Things When Change is Hard* by Chip and Dan Heath**

Switch was written to help people with implementation. If you have already learned the art of dealing with people, the next step is to find safe and effective ways to lead individuals and groups through the process of change. The authors take the reader through a thought-provoking analogy of the Rider, the Elephant, and the Path. The Rider is our intellect, the Elephant is our motivation, and the Path represents our circumstances when trying to change. All three must work together. Chip and Dan Heath provide practical steps that can help us implement change when change is hard. The book is important. It will help builders become adept at implementing change on projects. Without this practical knowledge, it can be easy to fall victim to circumstance.

Recommended Reading

***How to Stop Worrying and Start Living* by Dale Carnegie**

In *How to Stop Worrying and Start Living* Dale Carnegie again provides advice that has no equal in the form of how to reduce stress and worry. For those who tend to suffer from stress and anxiety, the practical steps he lists in the book can make the difference between a happy and productive life and one of constant misery.

***The Ideal Team Player: How to Recognize and Cultivate the Three Essential Virtues* by Patrick M. Lencioni**

The Ideal Team Player describes, in the form of a fictional story, the attributes needed to be a good team player. Patrick Lencioni demonstrates the need to be humble, hungry, and smart. You may think you know what those words mean, but there is much more to each of them than initially comes to mind. The message of the book is helpful, insightful, and life-changing if applied.

***The Five Dysfunctions of a Team: A Leadership Fable* by Patrick Lencioni**

This book is essential for anyone who is part of a team or leads a team. Every team will struggle if they do not apply the principles of building trust and employing healthy conflict. It's hard to believe, but the answer to how we can be a good team is simple and contained in this book.

The Speed of Trust: The One Thing that Changes Everything **by Stephen R. Covey, et al.**

Have you ever wondered what it takes for someone to trust you? Would you like to know? Character and competence. A person must know your intentions, character, track record, and abilities. Stephen R. Covey outlines a pattern of behavior for us that we can use to build trust, which is the first step in team building.

Extreme Ownership: How U.S. Navy SEALs Lead and Win, **by Jocko Willink and Leif Babin**

If you have ever struggled with any form of a victim mentality, consider reading this book. Jocko Willink provides examples from his experience on deployment about how to be accountable, own the mission, and move forward as a competent team leader.

The Life-Changing Magic of Tidying Up: The Japanese Art of Decluttering and Organizing, **by Marie Kondo.**

In this book, Marie Kondo explains the Japanese philosophy of practicing cleanliness and organization in an effort to strive for perfection. It is hard for people to understand the need for cleanliness on projects in this industry. That is primarily because we are all conditioned for mediocrity. Kondo, with her fanatical approach to tidying, can inspire the reader to reach for perfection and joy in cleaning. Everything on-site should bring joy to the workers, owners, and management team. She will show you how.

Conclusion

In conclusion, I wish I could ask you if you enjoyed this work, but I'm not sure that is the point. My ultimate hope is that it makes you better, entices you to expect more, and pushes you out of your comfort zone. So, actually, I hope this has been a little bit difficult, and that is has challenged you a bit. Why? Because I care about your future, your family, and your enjoyment in life-and I know that those things only come from putting in the work, being dedicated to learning, and stretching ourselves to new vistas of achievement. My hope also is that builders can enjoy their work in construction. I can say that I enjoy my career, and it saddens me to hear of so many disliking their role and warning others to choose another path. So, if this work helps Supers to better enjoy their role, control their environments, and be happier and healthier at home, it has done its job.

I also understand the draft nature of this book. Again, my hope is that, through feedback, we can garner stories, visuals, and additional concepts from some of the best in our industry. If we do that, we will have a work that can be used to train the next generation of builders. To do this, I am asking you to share this book, to rate it highly, and to support our mission of respecting workers, training leaders, and preserving families.

I plan to also publish three more books titled *Elevating Construction Senior Superintendents, Elevating Construction Field Engineers*, and, *Elevating Construction Foremen*. To do this, I need this work to pave the way and do well, so, please rate, share, and support this book. If it has pushed you a little out of your comfort zone, left you wanting more, and encouraged a love of learning, then I feel we have done well. Please help me in elevating the entire construction industry from coast to coast.

If you have any need for support, please reach out to me at

Jasons@elevateconstructionist.com. Our company's mission is to help individuals and companies to take their next step.

You can also check us out on our podcast by searching the, "Elevate Construction Podcast." Until we meet again, On We Go!

The End

Made in the USA
Coppell, TX
17 January 2023

11241162R10144